Vibration Mechanics

# Vibration Mechanics

## Linear Discrete Systems

by

M. Del Pedro and P. Pahud
*Swiss Federal Institute of Technology,*
*Lausanne, Switzerland*

SPRINGER SCIENCE+BUSINESS MEDIA, B.V.

ISBN 978-94-010-5554-3 ISBN 978-94-011-3514-6 (eBook)
DOI 10.1007/978-94-011-3514-6

First edition Mécanique Vibratoire published in French
by Presses Polytechniques Romandes, 1989.
Translated from the French by
RICHARD ELLIS MA DPhil (Oxon)

---

*Printed on acid-free paper*


Originally published by Kluwer Academic Publishers in 1991
Softcover reprint of the hardcover 1st edition 1991

**TABLE OF CONTENTS**

# FOREWORD

## Objectives

This book is used to teach vibratory mechanics to undergraduate engineers at the Swiss Federal Institute of Technology of Lausanne. It is a basic course, at the level of the first university degree, necessary for the proper comprehension of the following disciplines.

- Vibrations of continuous linear systems (beams, plates)
- random vibration of linear systems
- vibrations of non-linear systems
- dynamics of structures
- experimental methods, rheological models, etc.

Effective teaching methods have been given the highest priority. Thus the book covers basic theories of vibratory mechanics in an appropriately rigorous and complete way, and is illustrated by numerous applied examples. In addition to university students, it is suitable for industrial engineers who want to strengthen or complete their training. It has been written so that someone working alone should find it easy to read.

## Description

The subject of the book is the vibrations of linear mechanical systems having only a finite number of degrees of freedom (ie discrete linear systems). These can be divided into the following two categories :

- systems of solids which are considered to be rigid, and which are acted upon by elastic forces and by linear resistive forces (viscous damping forces).
- deformable continuous systems which have been made discrete. In other words, systems which are replaced (approximately) by systems having only a limited number of degrees of freedom, using digital or experimental methods.

The behaviour of an elementary oscillator, which has one degree of freedom only, is first studied in a detailed way. In effect, a deep knowledge of this behaviour is indispensible to the proper understanding of complex systems. The free, steady and forced states are established by taking into consideration sub-critical, critical and super-critical damping for each case. The power consumed by an oscillator is analyzed with care, and an original energetic interpretation of the indicial (ie unit step) and unit impulse responses is proposed.

Two concepts are introduced : that of frequency response, which is most frequently used in vibratory mechanics, and that of admittance, or transfer function, which is usually used in control theory. The electrical analogue of force-current and the concept of circuits of force are briefly considered.

The oscillator with two degrees of freedom is then considered, emphasis being put on the conservative free state and the analysis of elastic coupling. Furthermore, the Frahm damper, a classical example of the steady state, is the subject of an optimization study.

The concept of a generalized linear discrete oscillator, having any finite number of degrees of freedom $n$ , is introduced on the basis of quadratic forms of the kinetic energy, of the potential energy and of the total power consumed. The methods of Lagrange and of Hamilton are used to establish the second order linear differential system, having $n$ variables. The solutions are established and analyzed in a systematic way, by order of increasing complexity.

That is to say, the null, proportional (ie the hypothesis of Caughey), and other dampings are considered in turn. The concepts of modal basis, and of real and complex modes, which are analysed in a rigorous manner, are made as natural as possible on the physical level. An original example of the visualization of complex modes is dealt with completely. Finally, the book finishes with an exposition of the principles of experimental modal analysis.

**Acknowledgements**

The authors extend thanks to Christine Benoit et Claudine Candaux for typing the manuscript, to José Dias Rodrigues for his contribution to the last chapter and to Jean-François Casteu for the preparation of the figures.

They particularly thank Martin Schmidt who reviewed the text and corrected the proofs.

Lausanne, march 1988
Pierre PAHUD and Michel DEL PEDRO

# CHAPTER 1 INTRODUCTION

## 1.1 Brief history

Vibratory phenomena play a critical role in nearly all the branches of physics; including mechanics, electricity, optics, acoustics. In spite of their diversity, they are governed in all cases in the linear region, by the same laws of behaviour and can be studied by means of the same mathematical tools.

Man interested himself in vibrating phenomenon when he built the first musical instruments.

Musicians and philosophers seek the laws of the production of sound and apply them to the construction of musical instruments. For example, Pythagoras (582-507 BC) showed experimentally that if two strings are under identical tension, then their tones will differ by one octave when the length of one is double that of the other.

Despite the knowledge acquired by the Ancients, it was necessary to wait until the beginning of the 17th century for Galileo (1564-1642) to show that the pitch of a sound is determined by the frequency of the vibrations. The phenomenon of beats was demonstrated by Sauver (1653-1716) at the end of the same century. It was Bruck Taylor (1685-1731) who, for the first time, rediscovered mathematically the experimental results of Galileo and other researchers.

Several renowned mathematicians have studied the problem of a vibrating string, for example, D. Bernoulli (1700-1782), d'Alembert (1717-1783), Euler (1707-1783), Lagrange (1736-1813) and Fourier (1768-1830).

Their studies have shown that a string can vibrate laterally in different ways, called modes of vibration. The first mode corresponds to the lowest frequency.

The displacement of the string is semi-sinusoidal. The second mode has twice the frequency of the first and the displacement of the string is sinusoidal, which produces a **node** at its centre.

Sauveur gave the name **fundamental** to the lowest frequency and **harmonics** to the higher frequencies.

The linear superposition of the harmonics was proposed for the first time by Bernouilli. Finally in 1822, Fourier presented his celebrated paper on the theory of harmonic series.

In about 1750, d'Alembert established the differential equation governing the vibrations of a string. The wave characteristics of this equation were recognized later and this led to it being called the **wave equation**.

As a result of Hooke's law (stated in 1676), Euler and Bernouilli studied the vibrations of beams. Their calculations were based on the conservation of energy. This method was developed later by Lord Rayleigh (1842-1919) and carries his name. The study of the vibrations of plates and membranes was undertaken much later, in particular by Kirchoff (1842-1887) and Poisson (1781-1840).

Among contemporary researchers, we will mention Stodola (1859-1943) who established a method for the analysis of the vibrations of beams at the time of his work on the vibrations of turbine blades.

During the course of recent decades, the rapid development of computers and of experimental methods has resulted in important progress in vibratory mechanics. It is now possible to undertake the study of complex systems, subject to any deterministic or random stimulations.

As we have said, the first vibratory phenomena studied concerned the creation and transmission of sound. People later became interested in the vibrations of mechanical systems, but it was with the adoption of alternating current, as a means of transporting energy, that the study of oscillating phenomena received increased attention.

Consequently it is not astonishing that it is the electrical engineers who, having taken over from the mechanical engineers and physicists, have developed the most convenient and most profitable state-of-the art methods of calculation. These methods have been transformed since the study of acoustical and mechanical vibrations. In this way there emerged progressively a general theory of vibrations, independent of their physical form. Furthermore, vibrations often pose problems of stability, which can be solved by means of the classical criteria of control theory.

## 1.2 Disruptive or useful vibrations

For a very long time, one has studied the vibrations of machines and structures almost uniquely with the objective of attenuating them and, if possible, eliminating them. This preoccupation is still essential but is no longer the only one. More and more nowadays one constructs machines and equipment which use vibratory mechanics to fulfil their desired function.

Here are several examples where the vibrations are a disruptive element and must be counteracted :

- The vibrations of machines, or of certain components of machines, are a cause of loss of precision, of noise, of premature wear and tear, and of fatigue. One means by fatigue of the materials a process of creation and development of cracks, leading finally to the complete rupture of the part. Moreover, vibrations cause a dissipation of energy, which lowers the efficiency because of the existence of passive resistances.

- The vibrations of cars, planes, trains and ships cause, in addition to the above drawbacks, the discomfort of the travellers and sometimes seriously reduce the safety of these vehicles. The shimmy of cars, the yawing of locomotives, the pitching of ships and the vibratory instability of aeroplane wings belong to this type of phenomena.

- The vibrations of large metal structures can become catastrophic in certain cases. Let us recall the collapse of the San Francisco suspension bridge brought about by resonant oscillations due to the wind.

One can say, on the subject of disruptive vibrations, that all primary motion is a source of vibrations. It is remarkable to note that a slight constructive modification can bring about a noticeable reduction or the complete elimination of these vibrations. That does not mean that it is always easy to correct a machine or an existing structure, but rather that shows the necessity of being concerned about vibrations in every construction project. The problem can be summed up as follows : discover the sources of vibrations, study the transmission of the vibrations to the rest of the construction, seek the possibilities for resonance then imagine the means to attenuate the phenomenon.

On the contrary, in machines and equipment using vibrations it is a matter of optimizing the efficiency of the source of these vibrations. It is equally necessary to choose the type of vibrations best suited for the desired function and to assure that the transmission system does not suffer excessive fatigue. Here are several examples of this type of machine and equipment :

- Tamping and levelling machines for the automatic maintenance of railway track,
- all types of mechanical and ultrasonic vibrators,
- vibration transmission systems,
- vibrating polishers,
- lithotrite (medical apparatus to fragment kidney stones).

# CHAPTER 2 THE LINEAR ELEMENTARY OSCILLATOR OF MECHANICS

## 2.1 Definitions and notation

A **linear elementary oscillator** is a mechanical system with one degree of freedom whose behaviour as a function of time is described by a linear second order differential equation with constant coefficients.

Let us recall that a mechanical system possesses only one degree of freedom when its configuration can be, at each moment, characterized by a single variable. More generally, a system has $n$ degrees of freedom if the minimum number of variables necessary to define its configuration is equal to $n$ .

The system is qualified as linear when it can be described by means of linear differential equations. An elementary mechanical oscillator has the elements shown in figure 2.1, that is to say :

- a rigid **mass** $m$,
- a massless **spring** which supplies an elastic force proportional and opposed to the displacement $x(t)$ ; the coefficient of proportionality $k$ is called the **stiffness** - or **rigidity** - of the spring;
- a **damper** which supplies a resistive force, proportional and opposed to the velocity $\dot{x}(t)$ , the coefficient of proportionality $c$ is called the **linear viscous damping factor** or, more simply, the **resistance** of the system.

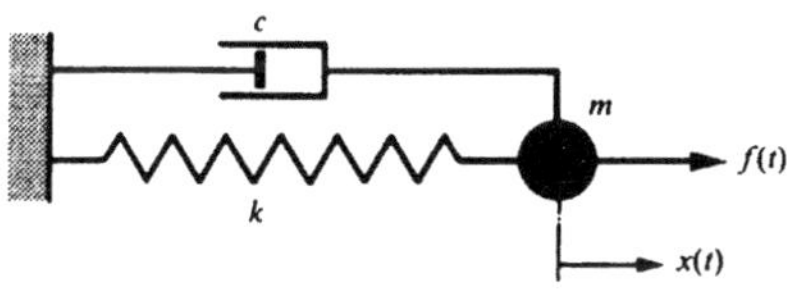


Fig. 2.1 Standard diagram of a linear elementary mechanical oscillator.

## 2.2 Equation of motion and vibratory states

If an external force $f(t)$ acts on a mass, Newton's law is written :

$$m\ddot{x} = -kx - c\dot{x} + f(t)$$

which gives $$m\ddot{x} + c\dot{x} + kx = f(t) \qquad (2.1)$$

It is clearly a linear second order differential equation with constant coefficients, thus it is worth studying the solutions in detail for the following reasons :

- numerous practical systems can be described by an equation of this type;
- a good understanding of the elementary oscillator makes it easier to study more complex systems which will be undertaken later.

Let us first give several definitions relative to the principal types of behaviour of the system.

- The **free state** corresponds to the general solution of the differential equation without the right-hand side, that is to say for $f(t) = 0$.
- The **forced state** corresponds to the complete solution with the right-hand side. It therefore depends on the nature of $f(t)$; unit-impulse force, harmonic, any periodic form, random, etc.
- The **steady state** is the forced state, after the disappearance of transient terms, caused by a periodic force. It is not influenced by the initial conditions. When the system is conservative, that is to say when the damping is zero, the steady state does not exist, strictly speaking, for the initial conditions indefinitely influence the behaviour of the system.

The linear elementary oscillator, shown schematically in figure 2.1, is encountered in reality in extremely varied forms. We will give several examples below.

Very often, in practical problems, only small movements of a system can be described by a linear differential equation. This is a considerable limitation, and sometimes it is necessary to abandon the hypothesis of linearity in order to avoid larger errors which can remove all meaning from the results found. Thus one is led to study the **non-linear elementary oscillator**.

## 2.3 Modified forms of the equation of motion

Let us return to equation (2.1). It expresses the fact that the external force $f(t)$ is equal to the sum of the three internal forces of the system, that is to say, the force of inertia $m\ddot{x}$ the force of viscous resistance $c\dot{x}$ and the elastic force $kx$ .

$$m\,\ddot{x} + c\,\dot{x} + k\,x = f(t)$$

Let us divide by the mass

$$\ddot{x} + 2\,\frac{c}{2m}\,\dot{x} + \frac{k}{m}\,x = \frac{1}{m}\,f(t)$$

and introduce the notation :

$\omega_0^2 = \frac{k}{m} \Rightarrow \omega_0 = \sqrt{\frac{k}{m}}$ **natural angular frequency** of a conservative system (2.2)

$\lambda = \frac{c}{2m}$ **damping coefficient** (2.3)

$\eta = \frac{c}{2\,m\,\omega_0} = \frac{\lambda}{\omega_0}$ **damping factor** (or **relative damping**) (2.4)

Consequently, the differential equation is written :

$$\ddot{x} + 2\,\lambda\,\dot{x} + \omega_0^2\,x = \frac{1}{m}\,f(t) \tag{2.5}$$

The four terms have the physical dimension of acceleration.

If one now divides (2.1) by the stiffness $k$ , the equation is made up of terms having the dimension of displacement :

$$\frac{m}{k}\,\ddot{x} + \frac{c}{k}\,\dot{x} + x = \frac{1}{k}\,f(t)$$

The right-hand side represents the **elastic displacement** which would give rise to the external force if the system was only made up of the stiffness $k$

$$x_e(t) = \frac{1}{k}\,f(t) \tag{2.6}$$

By using the above definitions, the equation becomes

$$\frac{1}{\omega_0^2}\,\ddot{x} + \frac{2\lambda}{\omega_0^2}\,\dot{x} + x = x_e(t) \tag{2.7}$$

This last form of the equation of motion is very convenient when one investigates the forced state caused by a displacement imposed on the system.

# CHAPTER 3 THE FREE STATE OF THE ELEMENTARY OSCILLATOR

## 3.1 Conservative free state · Harmonic oscillator

The free state describes the behaviour of an elementary oscillator after it has been released from **initial conditions,** without the subsequent supply of energy by an external force. That is, when $f(t) = 0$.

This release is defined, at time $t = 0$, by an initial displacement $X_0 = x(0)$ .

The oscillator is said to be conservative when the damping is zero : $c = 0 \Rightarrow \lambda = 0$ . In this way equation (2.5) becomes

$$\ddot{x} + \omega_0^2 x = 0 \tag{3.1}$$

The solution of this equation gives the displacement of the mass

$$x = A \cos \omega_0 t + B \sin \omega_0 t = A \cos \omega_0 t + B \cos(\omega_0 t - \pi/2) \tag{3.2}$$

The two harmonic functions are equal to the projections, onto an axis, of two vectors, of length $A$ and $B$ , rotating at the same angular velocity $\omega_0$ (figure 3.1). Their sum is therefore equal to the projection of the resulting vector of length $X$ and the phase $\varphi$.

$$x = X \cos(\omega_0 t - \varphi) \tag{3.3}$$

The new constants of integration $X$ and $\varphi$ are linked to the values of $A$ and $B$ by the obvious relations

$$\begin{cases} X = \sqrt{A^2 + B^2} \\ \mathrm{tg}\varphi = \dfrac{B}{A} \end{cases} \tag{3.4}$$

The displacement of the mass is thus an harmonic motion of angular frequency $\omega_0$ , of frequency $f_0$ and of period $T_0$ . The system is known as the **harmonic oscillator**.

$$f_0 = \frac{1}{2\pi}\,\omega_0 \tag{3.5}$$

$$T_0 = \frac{1}{f_0} = \frac{2\pi}{\omega_0} \tag{3.6}$$

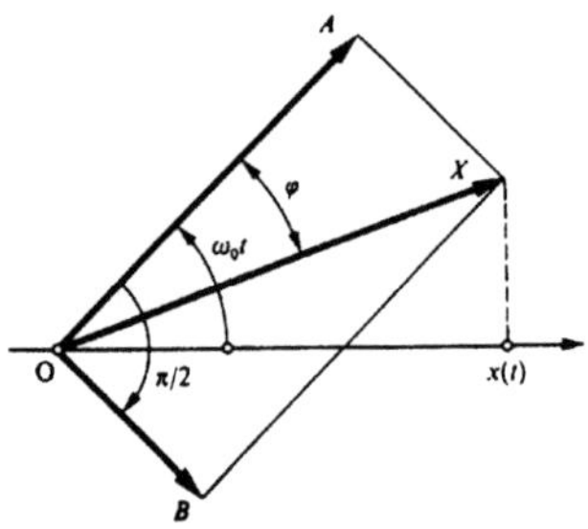


Fig. 3.1 Rotating vectors representing the displacement x(t)

One obtains the velocity by differentiation of (3.3)

$$\dot{x} = -\,\omega_0\,X\,\sin(\omega_0 t-\varphi) = \omega_0\,X\,\cos(\omega_0 t-\varphi+\pi/2) \tag{3.7}$$

A second differentiation gives the acceleration

$$\ddot{x} = -\,\omega_0^2\,X\,\cos(\omega_0 t-\varphi) = \omega_0^2\,X\,\cos(\omega_0 t-\varphi+\pi) \tag{3.8}$$

These results mean that the velocity and acceleration are 90° and 180° respectively out of phase with the displacement (figures 3.2 and 3.3).

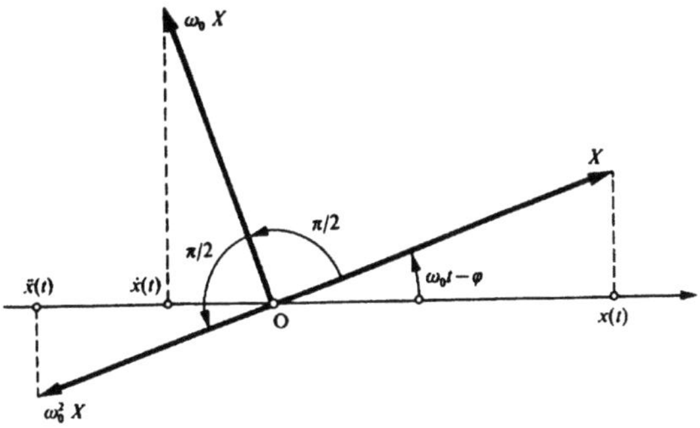


Fig. 3.2 Rotating vectors representing the displacement, the velocity and acceleration

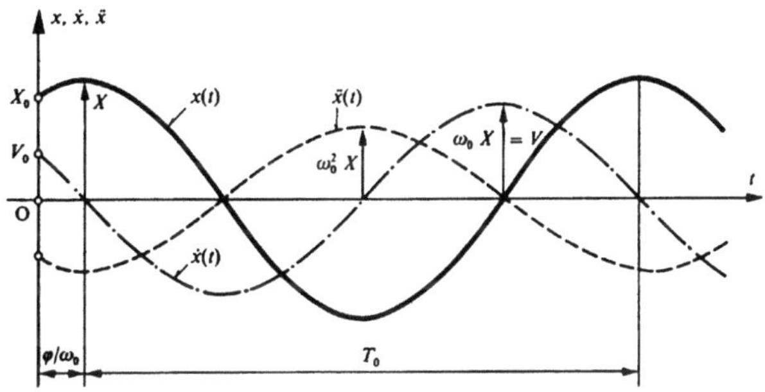


Fig. 3.3 Displacement x, velocity $\dot{x}$ and acceleration $\ddot{x}$ for the conservative free state (harmonic oscillator)

## 3.2 Conservation of energy

Always in the absence of damping, return to the free state equation

$$m \frac{d^2x}{dt^2} + k\, x = 0$$

By introducing the variable for the velocity of the mass

$$v = \frac{dx}{dt}$$

the acceleration takes the form

$$\frac{d^2x}{dt^2} = \frac{dv}{dt} = \frac{dv}{dx} \cdot \frac{dx}{dt} = \frac{dv}{dx} v$$

The equation becomes

$$m v \frac{dv}{dx} + k x = 0$$

which gives finally

$$m v dv + k x dx = 0$$

One obtains by integration

$$\frac{1}{2} m v^2 + \frac{1}{2} k x^2 = T + V = H = \text{const.} \tag{3.9}$$

This result shows that the total mechanical energy $H$ of the system, which is equal to the sum of the kinetic energy $T$ and the potential energy $V$, is constant - is conserved - in the free state of an undamped oscillator.

When the velocity is zero, the displacement has its maximum value $X$ and all the energy is in the form of potential energy. Conversely, when the displacement is zero, the velocity reaches its maximum $V = \omega_0 X$ and all the energy is kinetic. Thus one has

$$H = \frac{1}{2} m (\omega_0 X)^2 = \frac{1}{2} k X^2$$

Relation (3.9) can be used to establish the differential equation of motion of a conservative system. In effect, by differentiating this relation with respect to time, one can write

$$\frac{dH}{dt} = 0 \tag{3.10}$$

## 3.3 Examples of conservative oscillators

### 3.3.1 Introduction

The elementary oscillator, as we have previously called it, is encountered in practice in very diverse forms. Moreover, the assumption that a system is linear and has only one degree of freedom results nearly always in approximations, sometimes radical, which it is necessary to be aware of when interpreting the results. We are going to deal with several applied examples.

### 3.3.2 Mass at the end of a wire

Find the natural frequency of the system shown in figure 3.4 by assuming that the wire has a fixed length and that the pulleys are massless (in other words, their inertias of rotation and of translation are zero).

Let us adopt the notation

| | |
|---|---|
| $x_0$, $x_{10}$, $x_{20}$, $T_0$ | displacements and static tensions due to gravity |
| $x$ , $x_1$ , $x_2$ , $T$ | displacements and dynamic tensions about the equilibrium positions |
| $x'$, $x_1'$ , $x_2'$ , $T'$ | total displacments and tension |

For this first problem, let us proceed systematically by finding first of all the static position of the mass. The static equilibrium of the mass and of the pulleys gives

$$mg = T_0$$
$$2\ T_0 = k_1\ x_{10}$$
$$2\ T_0 = k_2\ x_{20}$$

Moreover, the wire having a fixed length, one can write

$$x_0 = 2\ (x_{10} + x_{20})$$

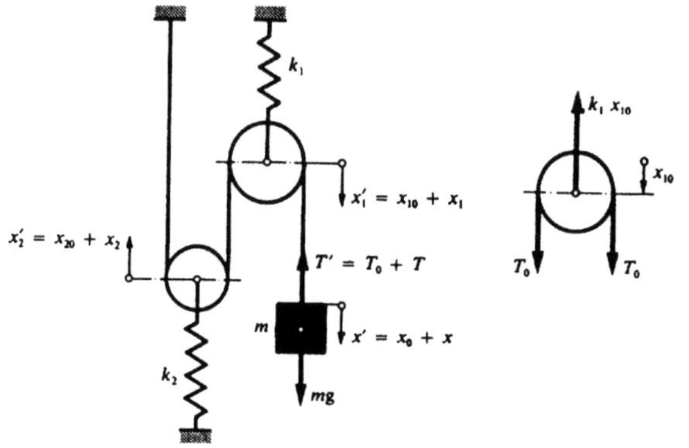


Fig. 3.4 Oscillator consisting of wire of fixed length and two massless pulleys

whence, by eliminating $x_{10}$ and $x_{20}$

$$x_0 = T_0 \cdot 4 \left(\frac{1}{k_1} + \frac{1}{k_2}\right)$$

One introduces the equivalent stiffness $k_e$

$$x_0 = \frac{T_0}{k_e}$$

$$\frac{1}{k_e} = 4 \left(\frac{1}{k_1} + \frac{1}{k_2}\right) \tag{3.11}$$

In summary

$$k_e x_0 - mg = 0 \tag{3.12}$$

Let us now write Newton's law for the total displacement of the mass

$$m \ddot{x}' = mg - T' \tag{3.13}$$

One has moreover

$$2 T' = k_1 x_1' = k_2 x_2' \tag{3.14}$$

$$x' = 2 (x_1' + x_2') \tag{3.15}$$

and consequently

$$x' = T' \cdot 4\left(\frac{1}{k_1} + \frac{1}{k_2}\right) = \frac{T'}{k_e} \quad => \quad T' = k_e\, x'$$

Equation (3.13) becomes

$$m\,\ddot{x}' = mg - k_e\, x'$$

then, by replacing $x'$ by $x_0 + x$

$$m\,(0 + \ddot{x}) = mg - k_e(x_0 + x)$$

or again

$$m\,\ddot{x} + k_e\, x = mg - k_e\, x_0$$

The right-hand side being zero according to (3.12), one has finally

$$m\,\ddot{x} + k_e\, x = 0 \qquad (3.16)$$

Before continuing, we are going to re-derive this result by setting the derivative of the energy of the system equal to zero in accordance with (3.10) :

- **kinetic energy**

$$T = \frac{1}{2}\, m\, \dot{x}'^2$$

- **potential energy**

$$V = \frac{1}{2}\, k_1\, x_1'^2 + \frac{1}{2}\, k_2\, x_2'^2 - m\, g\, x'$$

Relations (3.14) and (3.15) give

$$x'_1 = \frac{k_2}{2(k_1+k_2)} x' \qquad\qquad x'_2 = \frac{k_1}{2(k_1+k_2)} x'$$

The potential energy is written as a function of $x'$ only, as follows :

$$V = \frac{1}{2} \frac{k_1 k_2}{4(k_1+k_2)} x'^2 - mg\, x'$$

then, by introducing the equivalent stiffness $k_e$

$$V = \frac{1}{2} k_e x'^2 - mg\, x'$$

The sum of the kinetic and potential energies is thus :

$$H = T + V = \frac{1}{2} m \dot{x}'^2 + \frac{1}{2} k_e x'^2 - mg\, x' \tag{3.17}$$

The condition $\frac{dH}{dt} = 0$ becomes in this particular case

$$\dot{x}' (m \ddot{x}' + k_e x' - mg) = 0 \tag{3.18}$$

The solution $\dot{x}' = 0$ restores the static equilibrium; in effect

$$\dot{x}' = 0 \Rightarrow x' = \text{const.} = x_0 + x(t) \Rightarrow \begin{cases} x(t) = 0 \\ x' = x_0 \end{cases}$$

By setting the term in parenthesis in equation (3.18) to zero and by replacing $x'$ by $x_0 + x$, one clearly re-derives the equation of motion (3.16).

The natural frequency for small movements of the system is thus, using (3.5) and (2.2) :

$$f_0 = \frac{1}{2\pi} \omega_0 = \frac{1}{2\pi} \sqrt{\frac{k_e}{m}}$$

Numerical example

Let us choose the following values for the constants

$$m = 50 \text{ kg} \quad k_1 = 3 \cdot 10^4 \text{ N/m} \quad k_2 = 5 \cdot 10^4 \text{ N/m} \quad g = 9.81 \text{ m/s}^2$$

This gives

$$k_e = \frac{k_1 \cdot k_2}{4(k_1+k_2)} = \frac{3 \cdot 5}{4(3+5)} 10^4 = 4690 \text{ N/m} \tag{3.11}$$

$$x_0 = \frac{mg}{k_e} = \frac{50 \cdot 9.81}{4690} = 0.105 \text{ m}$$

$$f_0 = \frac{1}{2\pi} \left(\frac{4690}{50}\right)^{1/2} = 1.54 \text{ Hz}$$

Comments

- The weight mg of the mass does not play a role in the natural frequency of the system, which one can assert a priori for a problem of this type. It is therefore possible to proceed more rapidly by considering only the dynamic displacements, that is to say :

$$\left.\begin{array}{l} 2\,T = k_1\, x_1 \\ 2\,T = k_2\, x_2 \\ \;\; x = 2\,(x_1+x_2) \end{array}\right\} \Rightarrow \quad x = T \cdot 4 \left(\frac{1}{k_1} + \frac{1}{k_2}\right) = \frac{T}{k_e}$$

$$m\ddot{x} = mg - (T_0+T) = (mg-T_0) - T \quad \Rightarrow \quad m\ddot{x} + T = 0$$

  One re-derives equation (3.16).

- In order that the oscillations stay in the linear region, it is necessary that the wire stays taut, which limits the amplitude of motion : $X < x_0 = 0.105$ m.

· The system will have three degrees of freedom, instead of only one, if the mass of the pulleys is taken into account. The fundamental frequency of such a system, that is to say the lowest natural frequency, will be less than 1.54 Hz.

### 3.3.3 Lateral vibrations of a shaft

A cylindrical steel shaft with a circular cross-section carries a disk of relatively large diameter but insignificant thickness (fig. 3.5). By assuming the disk is a point mass m and by neglecting the mass of the shaft, let us calculate the natural frequency of the vibrations of bending of the system.

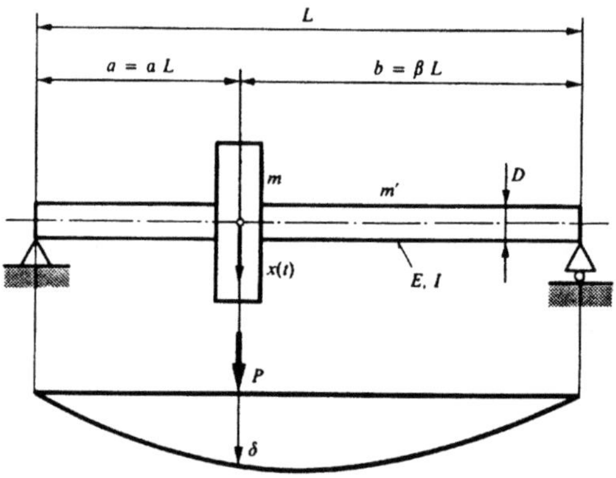


Fig. 3.5 Vibrations of a machine shaft

Within the framework of these assumptions (point mass, massless shaft) the system is an elementary oscillator. The small movements of the mass are governed by the equation.

$$m \ddot{x} + k x = 0 \quad \Rightarrow \quad f_0 = \frac{1}{2\pi}\sqrt{\frac{k}{m}}$$

One determines the stiffness $k$ by supposing that a static force $P$ is applied to the mass. This force causes a displacement $\delta$ having the value

$$\delta = P \frac{a^2 b^2}{3\, L\, E\, I}$$

In this expression, $E$ and $I$ are respectively the modulus of elasticity and the bending moment of inertia. From the definition of the stiffness

$$P = k\,\delta \quad \Rightarrow \quad k = \frac{P}{\delta} = \frac{3\, L\, E\, I}{a^2 b^2} = \frac{3\, E\, I}{\alpha^2 \beta^2 L^3}$$

The natural frequency of the system is thus

$$f_0 = \frac{1}{2\pi} \cdot \frac{1}{\alpha\, \beta\, L} \sqrt{\frac{3\, E\, I}{m\, L}}$$

Numerical example

$L = 1$ m $\quad a = 0.4$ m $\quad b = 0.6$ m $\quad \Rightarrow \quad \alpha = 0.4 \quad \beta = 0.6$

$D = 12$ cm $\quad m = 300$ kg $\quad E = 2.1 \cdot 10^{11}$ N/m$^2$ $\quad \varrho = 7.8 \cdot 10^3$ kg/m$^3$

$$I = \frac{\pi\, D^4}{64} = \frac{\pi \cdot 12^4}{64} = 1018 \text{ cm}^4 = 1.018 \cdot 10^{-5} \text{ m}^4$$

$$f_0 = \frac{1}{2\pi} \cdot \frac{1}{0.4 \cdot 0.6 \cdot 1} \left( \frac{3 \cdot 2.1 \cdot 1.018}{3 \cdot 1} \; 10^{11-5-2} \right)^{1/2} = 97 \text{ Hz}$$

Comments

- The vibrations of the shafts of machines will be considered later. We will see that such systems possess an infinite number of natural frequencies, the lowest of which is called the fundamental frequency.

- The fact that one has neglected the mass of the shaft $m'$ results in one over estimating the natural frequency of the system. One can obtain a lower bound by replacing $m$ by (m + m') in the expression for the natural frequency.

$$m' = \frac{\pi D^2}{4} L \rho = 88{,}2 \text{ kg} \quad \Rightarrow \quad m + m' = 388 \text{ kg} \quad \Rightarrow \quad f_0' = 85 \text{ Hz}$$

The actual fundamental frequency, calculated by means of a finite element program, is 90.2 Hz. It is thus well bounded by the two preceding values : 85 < 90.2 < 97.

### 3.3.4 Pendulum system

The pendulum system shown in figure 3.6 rolls without sliding on a horizontal plane. It consists of a cylinder of mass M , of moment of inertia J , joined together rigidly by a rod to an assumed point mass m . The mass of the rod being negligible, let us establish, by differentiating the mechanical energy, the differential equation for small movements about the equilibrium position.

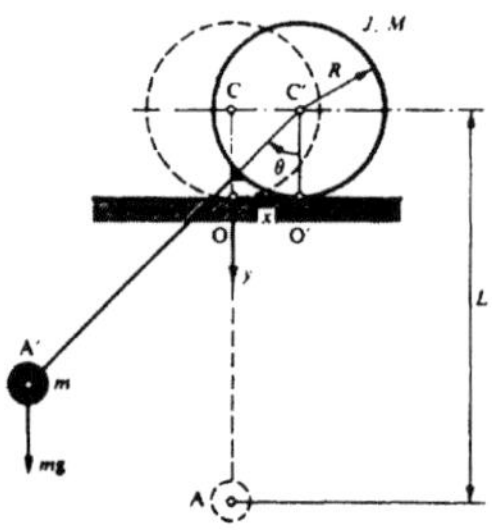


Fig. 3.6 Pendulum system rolling on an horizontal plane

In the inertial reference frame Oxy, the centre C' is marked out by its radius-vector $\vec{OC'}$ :

$$\vec{OC'} = \begin{Bmatrix} R\,\theta \\ -R \end{Bmatrix}$$

The velocity of C' is obtained by differentiation, whence

$$\vec{V}_{C'} = \begin{Bmatrix} R\dot{\theta} \\ 0 \end{Bmatrix}$$

Doing the same for A' one obtains

$$\vec{OA'} = \begin{Bmatrix} R\theta - L\sin\theta \\ -R + L\cos\theta \end{Bmatrix} \quad => \quad \vec{V}_{A'} = \begin{Bmatrix} (R - L\cos\theta)\,\dot{\theta} \\ -L\sin\theta \cdot \dot{\theta} \end{Bmatrix}$$

Knowing the vector expression for the velocity, one can calculate the kinetic energy of the system

$$T = \frac{1}{2} J\dot{\theta}^2 + \frac{1}{2} M\,|\vec{V}_{C'}|^2 + \frac{1}{2} m\,|\vec{V}_{A'}|^2$$

$$T = \frac{1}{2}\dot{\theta}^2 \left(J + M R^2 + m\,(R^2 + L^2 - 2 R L \cos\theta)\right) \tag{3.19}$$

The variation of the potential energy is due to the vertical displacement of the point mass m only

$$V = m\,g\,L\,(1 - \cos\theta) \tag{3.20}$$

From relation (3.10), one obtains the differential equation of the system by differentiating the total energy :

$$\frac{dH}{dt} = \frac{d}{dt}(T + V)$$

which gives from (3.19) and (3.20)

$$0 = \dot{\theta}\,\ddot{\theta}\left(J + M R^2 + m\,(R^2 + L^2 - 2 R L \cos\theta)\right) + m R L \dot{\theta}^3 \sin\theta + m g L \dot{\theta} \sin\theta$$

Only the solution $\dot{\theta} \neq 0$ , interests us, which gives, after division by $\dot{\theta}$

$$(J + M R^2 + m (R^2 + L^2 - 2 R L \cos\theta)) \ddot{\theta} + m (R L \dot{\theta}^2 + g L) \sin\theta = 0$$

For small movements, the simplifications $\sin\theta \approx \theta$ , $\cos\theta \approx 1$ and $\dot{\theta}^2 \ll g/R$ lead to the linear differential equation

$$(J + M R^2 + m (R - L)^2) \ddot{\theta} + m g L \theta = 0 \tag{3.21}$$

This is clearly the equation for a conservative elementary oscillator. Its natural frequency has the value

$$f_0 = \frac{1}{2\pi} \sqrt{\frac{m g L}{J + M R^2 + m (R - L)^2}} \tag{3.22}$$

One easily re-derives equation (3.21) by means of Lagrange's differential equation

$$\frac{d}{dt} \frac{\partial T}{\partial \dot{\theta}} - \frac{\partial T}{\partial \theta} + \frac{\partial V}{\partial \theta} = 0$$

3.3.5 Helmholtz resonator

Calculate the natural frequency of the oscillations of a column of gas contained in a tube whose extremities come out into an unbounded medium at constant pressure and into a rigid container respectively (figure 3.7).

$p_0, \varrho_0$ pressure and density of the gas at the end C (undefined medium)

| | |
|---|---|
| $p$ , $\varrho$ | pressure and density of the gas at the end B (container) |
| V | volume of the container |
| L , A | length and cross-section of the tube |

One such system is an elementary oscillator, known as the **Helmholtz resonator**, if one makes the following assumptions :

- the variation of pressure is much smaller than the mean pressure : $|p_0 - p| \ll p_0$
- the volume of the tube is much less than that of the container $L A \ll V$

In these conditions, the mass of the column of gas can be considered constant

$$m = A L \varrho_0$$

Let us write Newton's law for the displacment x(t) of this column

$$A L \varrho_0 \ddot{x} = - A(p - p_0)$$

$$\ddot{x} + \frac{p - p_0}{L \varrho_0} = 0 \qquad (3.23)$$

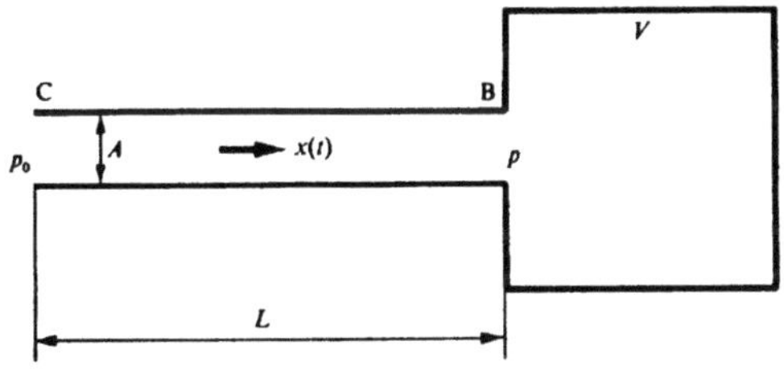


Fig. 3.7 Helmholtz resonator

In order to establish a relation between x and p, one assumes an isentropic behaviour for the gas, that is to say $p \cdot \rho^{-\gamma} = \text{constant}$, $\gamma$ being the isentropic exponent.

It therefore gives by differentiating

$$dp = \frac{\gamma\, p}{\rho}\, d\rho$$

The variations being small in comparison to the mean values, one can assume

$$dp = \frac{\gamma\, p_0}{\rho_0}\, d\rho \qquad (3.24)$$

If $T_0$ is the absolute temperature and R the gas constant, one has

$$\frac{p_0}{\rho_0} = R\, T_0$$

and the relation (3.24) becomes

$$dp = \gamma\, R\, T_0\, d\rho \qquad (3.25)$$

The elementary mass which enters in the container has the value

$$dm = A\, \rho_0\, dx$$

It causes a specific variation of the mass

$$d\rho = \frac{dm}{V} = \frac{A\, \rho_0}{V}\, dx \qquad (3.26)$$

It gives, by eliminating $d\rho$ between (3.25) and (3.26),

$$dp = \frac{\gamma\, R\, T_0\, A\, \rho_0}{V}\, dx$$

then by integrating between 0 and t , with $x_0 = 0$

$$p - p_0 = \frac{\gamma R T_0 A \rho_0}{V} x \qquad (3.27)$$

By introducing this result into the relation (3.23), one obtains finally

$$\ddot{x} + \frac{\gamma R T_0 A}{L V} x = 0 \qquad (3.28)$$

One finds again the differential equation of an elementary oscillator with

$$\omega_0^2 = \frac{\gamma R T_0 A}{L V} \qquad f_0 = \frac{1}{2\pi} \sqrt{\frac{\gamma R T_0 A}{L V}} \qquad (3.29)$$

One knows that the speed of propagation of sound waves in a gas has the value

$$a = \sqrt{\gamma R T_0} \quad \Rightarrow \quad a^2 = \gamma R T_0 \qquad (3.30)$$

which enables one to write the preceding relations in the form

$$\omega_0^2 = \frac{a^2 A}{L V} \qquad f_0 = \frac{a}{2\pi} \sqrt{\frac{A}{L V}} \qquad (3.31)$$

Numerical example

$R = 287\ m^2/s^2\ °K$ $\qquad T_0 = 273 + 20 = 293\ °K$

$\gamma = 1.4$ (air) $\qquad L = 10\ m \quad A = 0.01\ m^2 \quad V = 5\ m^3$

$a = (1.4 \cdot 287 \cdot 293)^{1/2} = 343\ m/s$

$$f_0 = \frac{343}{2\pi} \left(\frac{0.01}{10 \cdot 5}\right)^{1/2} = 0.772\ Hz$$

## 3.4 Dissipative free state

The oscillator is described as **dissipative** when the damping is not zero. Let us return to the equation (2.5) with f(t) = 0.

$$\ddot{x} + 2\lambda\dot{x} + \omega_0^2 x = 0 \tag{3.32}$$

It has the solution

$$x = A\, e^{r_1 t} + B\, e^{r_2 t} \tag{3.33}$$

with

$$\begin{cases} r_1 = -\lambda + \sqrt{\lambda^2 - \omega_0^2} \\ r_2 = -\lambda - \sqrt{\lambda^2 - \omega_0^2} \end{cases} \tag{3.34}$$

It is necessary in the following text, to distinguish the following three cases, as a function of the value of the damping factor $\eta$ :

$\eta > 1$ super-critical damping
$\eta = 1$ critical damping
$\eta < 1$ sub-critical damping

Furthermore, it is convenient to introduce the quantity $\omega_1$ , which is always real and positive, and is defined as follows

$$\omega_1^2 = |\lambda^2 - \omega_0^2| = \omega_0^2\, |\eta^2 - 1| \tag{3.35}$$

### 3.4.1 Super-critical damping

When the damping factor is larger than unity, one must write, from the relation (3.35) above

$$\eta > 1 \;\Rightarrow\; \omega_1 = \sqrt{\lambda^2 - \omega^2} = \omega_0 \sqrt{\eta^2 - 1} \tag{3.36}$$

The roots (3.34) are then

$$\begin{cases} r_1 = -\lambda + \omega_1 \\ r_2 = -\lambda - \omega_1 \end{cases} \tag{3.37}$$

They are both negative and the displacement becomes

$$x = e^{-\lambda t} \, (A \, e^{\omega_1 t} + B \, e^{-\omega_1 t}) \tag{3.38}$$

One obtains the velocity by differentiation

$$\dot{x} = - e^{-\lambda t}((\lambda - \omega_1) \, A \, e^{\omega_1 t} + (\lambda + \omega_1) \, B \, e^{-\omega_1 t}) \tag{3.39}$$

When the oscillator is released with initial conditions corresponding to $x(0) = X_0$ , $\dot{x}(0) = V_0$ , the above relations give

$$\begin{cases} X_0 = A + B \\ V_0 = - (\lambda - \omega_1) \, A - (\lambda + \omega_1) \, B \end{cases}$$

One derives from these the constants of integration A and B

$$\begin{cases} A = \dfrac{1}{2\omega_1} \, (X_0 \, (\lambda + \omega_1) + V_0) \\ \\ B = \dfrac{-1}{2\omega_1} \, (X_0 \, (\lambda - \omega_1) + V_0) \end{cases} \tag{3.40}$$

Figure 3.8 shows $x(t)$ and $\dot{x}(t)$ in the case $X_0 > 0$ and $V_0 > 0$ ( $\Rightarrow A > 0$ , $B < 0$ ). These are aperiodic functions, the system is no longer an oscillator.

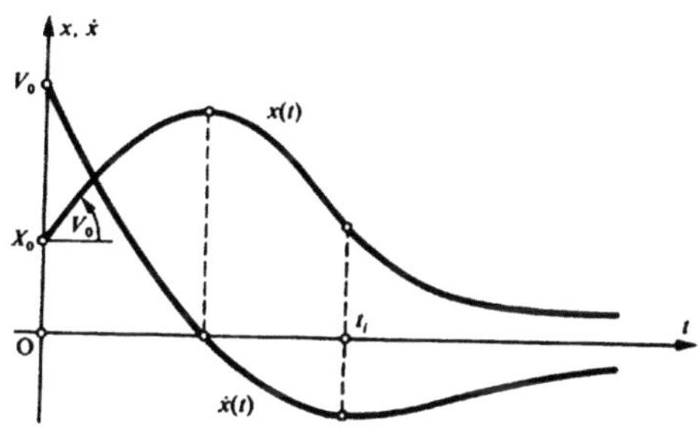


Fig. 3.8 Free state with super-critical damping

Taking account of (3.40) , the displacement x given by the relation (3.38) can be put in the form

$$x = e^{-\lambda t}\left(X_0 \operatorname{ch}\omega_1 t + \frac{\lambda X_0 + V_0}{\omega_1} \operatorname{sh}\omega_1 t\right) \qquad (3.41)$$

One easily confirms that $x(t)$ presents an inflexion point, and therefore $\dot{x}(t)$ an extreme value, for

$$t_i = \frac{1}{2\omega_1} \ln \frac{(\lambda+\omega_1)^2\,(X_0(\lambda-\omega_1) + V_0)}{(\lambda-\omega_1)^2\,(X_0(\lambda+\omega_1) + V_0)} \qquad (3.42)$$

The inflexion point would disappear if

$$\frac{(\lambda+\omega_1)^2}{(\lambda-\omega_1)^2} \cdot \frac{X_0(\lambda-\omega_1) + V_0}{X_0(\lambda+\omega_1) + V_0} < 1$$

in other words, when the initial velocity, for $X_0 > 0$ , is included in the interval

$$- X_0(\lambda+\omega_1) < V_0 < - X_0 \frac{\omega_0^2}{2\lambda} \qquad (= - X_0 \frac{k}{c})$$

The initial displacememt $X_0$ being fixed, figure 3.9 gives the shape of $x(t)$ for different values of the initial velocity.

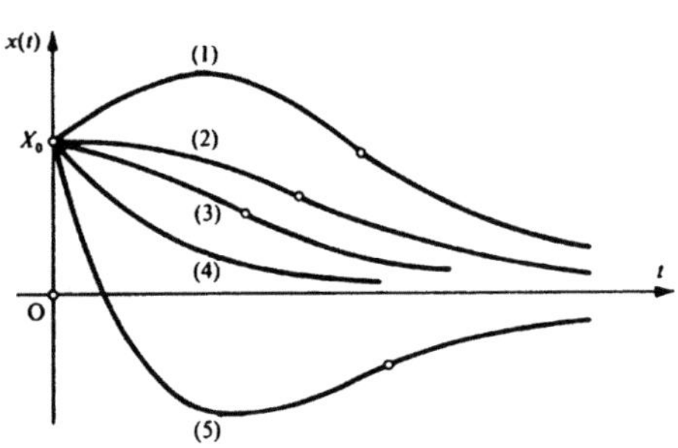


(1) $V_0 > 0$

(2) $V_0 = 0$

(3) $-X_0 \frac{\omega_0^2}{2\lambda} < V_0 < 0$

(4) $-X_0(\lambda+\omega_1) < V_0 < -X_0 \frac{\omega_0^2}{2\lambda}$

(5) $V_0 < -X_0(\lambda+\omega_1)$

· inflexion point

Fig. 3.9 Super-critical damping; behaviour of the system for different initial velocities

### 3.4.2 Critical damping

When $\eta = 1$ , the characteristic equation of (3.32) has a double solution

$$r_1 = r_2 = -\lambda = -\omega_0 \qquad (3.43)$$

The general solution of all second order differential equations is given by the linear combination of two specific linearly independent solutions; it is the same if $r_1 = r_2$.

A first specific solution is of the form

$$x_1 = e^{-\omega_0 t}$$

One can see, by substitution into the differential equation, that there exists a second specific solution of the form

$$x_2 = u(t)\, e^{-\omega_0 t}$$

In effect,

$$\dot{x}_2 = e^{-\omega_0 t}(\dot{u}(t) - \omega_0\, u(t))$$

$$\ddot{x}_2 = e^{-\omega_0 t}\, (\ddot{u}(t) - 2\omega_0\, \dot{u}(t) + \omega_0^2\, u(t))$$

With $\lambda = \omega_0$ , the substitution in (3.32) gives

$$\ddot{u}(t) = 0 \quad \Rightarrow \quad u(t) = Ct + D \quad \Rightarrow \quad x_2 = (Ct + D)e^{-\omega_0 t}$$

hence the general solution for $x(t)$

$$x = E\, x_1 + F\, x_2 = ((E + F\, D) + F\, C\, t)e^{-\omega_0 t}$$

and by changing the notation for the constants

$$x = (A + B\,t)e^{-\omega_0 t} \tag{3.44}$$

The velocity is thus

$$\dot{x} = ((B-\omega_0 A) - \omega_0\, B\, t)\, e^{-\omega_0 t} \tag{3.45}$$

The initial conditions $x(0) = X_0$ and $\dot{x}(0) = V_0$ determine the constants A and B

$$\begin{cases} A = X_0 \\ B = \omega_0\, X_0 + V_0 \end{cases} \tag{3.46}$$

and finally, by substitution into the preceding relations

$$x = (X_0 + (\omega_0\, X_0 + V_0)t)\, e^{-\omega_0 t} \tag{3.47}$$

$$\dot{x} = (V_0 - \omega_0(\omega_0\, X_0 + V_0)t)\, e^{-\omega_0 t} \tag{3.48}$$

These functions have the same shape as those of figure 3.8. The inflexion point has the abscissa

$$t_i = \frac{\omega_0\, X_0 + 2\, V_0}{\omega_0(\omega_0\, X_0 + V_0)} \tag{3.49}$$

This point does not exist if $t_i < 0$ , that is to say when the initial velocity, for $X_0 > 0$, is situated in the interval

$$-\,\omega_0\, X_0 < V_0 < -\,1/2\,\omega_0\, X_0 \tag{3.50}$$

### 3.4.3 Sub-critical damping

Let us return to the roots (3.34) of the characteristic equation

$$\begin{cases} r_1 = -\lambda + \sqrt{\lambda^2 - \omega_0^2} \\ r_2 = -\lambda - \sqrt{\lambda^2 - \omega_0^2} \end{cases}$$

They are complex when $\eta < 1 \Rightarrow \lambda^2 - \omega_0^2 < 0$ . The relation (3.35) allows one to write

$$\begin{cases} r_1 = -\lambda + j\,\omega_1 \\ r_2 = -\lambda - j\,\omega_1 \end{cases} \qquad (j = \sqrt{-1}\,) \tag{3.51}$$

In this case, the variable $\omega_1$ is the natural angular frequency of the damped oscillator

$$\omega_1 = \omega_0 \sqrt{1 - \eta^2} \tag{3.52}$$

and the displacement $x(t)$ becomes

$$x = e^{-\lambda t}(A \cos \omega_1 t + B \sin \omega_1 t) \tag{3.53}$$

One derives the velocity from it

$$\dot{x} = e^{-\lambda t}\,((-\lambda A + \omega_1 B) \cos \omega_1 t - (\omega_1 A + \lambda B) \sin \omega_1 t) \tag{3.54}$$

When the oscillator is released with any initial conditions $x(0) = X_0$, $\dot{x}(0) = V_0$, one finds the values for the constants

$$\begin{cases} A = X_0 \\ B = \dfrac{\lambda X_0 + V_0}{\omega_1} \end{cases} \tag{3.55}$$

The free state can be put in the form

$$x = e^{-\lambda t}(X_0 \cos \omega_1 t + \frac{\lambda X_0 + V_0}{\omega_1} \sin \omega_1 t) \tag{3.56}$$

which is the expression analogous to (3.41) in the sense that the hyperbolic functions are simply replaced by the trigonometric functions. These functions can be combined (see figure 3.1 page 10) and the displacement takes the simple form

$$x = X\, e^{-\lambda t} \cos(\omega_1 t - \varphi) \tag{3.57}$$

with

$$\left\{ \begin{array}{l} X = \sqrt{X_0^2 + (\frac{\lambda X_0 + V_0}{\omega_1})^2} \\ tg\varphi = \frac{\lambda X_0 + V_0}{\omega_1 X_0} \end{array} \right. \tag{3.58}$$

In this way, the quantity $X$ is always larger than the initial displacement $X_0$, except in the very specific case $V_0 = -\lambda X_0$. Likewise, the phase shift $\varphi$ is always different from zero, except if $V_0 = -\lambda X_0$.

As shown in figure 3.10, the function $x(t)$ is equal to the projection onto an axis of a rotating vector whose extremity traces a spiral.

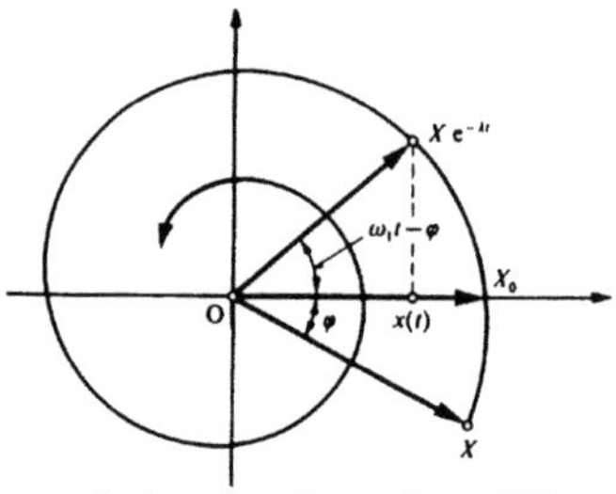


Fig. 3.10 Free state of a damped oscillator. Locus of the rotating vector

As a function of time, the displacement is represented by a damped sine wave enclosed between two envelopes $\pm X e^{-\lambda t}$ (figure 3.11).

The period of $x(t)$ - or more precisely the pseudo-period since the amplitude diminishes - has the value

$$T_1 = \frac{2\pi}{\omega_1} \tag{3.59}$$

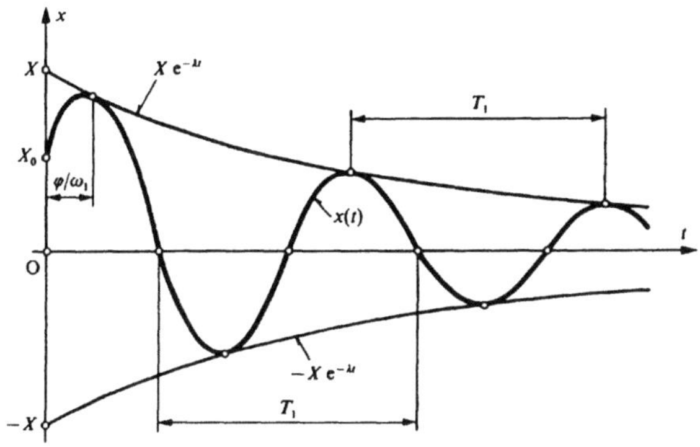


Fig. 3.11 Free state of the damped oscillator. Displacement as a function of time.

The damping reduces the angular frequency and increases the period of the oscillations. In effect, using again relation (3.52).

$$\omega_1 = \omega_0 \sqrt{1 - \eta^2} \qquad \omega_1 < \omega_0$$

$$T_1 = \frac{T_0}{\sqrt{1 - \eta^2}} \qquad T_1 > T_0 \tag{3.60}$$

In many practical problems, the term $\eta^2$ is very small with respect to 1 and the period $T_1$ can be confused with $T_0$ without appreciable error.

It is useful, in particular in the exploitation of measurements, to use the concept of **logarithmic decrement,** defined as follows

$$\Lambda = \frac{1}{n} \ln \frac{x(t)}{x(t + nT_1)} \tag{3.61}$$

It is thus the logarithm, divided by $n$, of the ratio of two displacements separated by a whole number $n$ of periods $T_1$ (more often than not, these displacements are the maxima). Using (3.57), it gives

$$x(t + nT_1) = X\, e^{-\lambda(t+nT_1)} \cos(\omega_1(t+nT_1) - \varphi) = e^{-n\lambda T_1}\, x(t)$$

whence

$$\Lambda = \frac{1}{n} \ell n\, e^{n\lambda T_1} = \lambda\, T_1 \qquad \text{(independent of n)}$$

Since

$$\lambda = \eta\, \omega_0 \quad \text{and} \quad T_1 = \frac{2\pi}{\omega_0 \sqrt{1 - \eta^2}}$$

the logarithmic decrement is only a function of the damping factor

$$\Lambda = \frac{2\,\pi\,\eta}{\sqrt{1 - \eta^2}} \tag{3.62}$$

and vice versa

$$\eta = \frac{\Lambda}{\sqrt{4\,\pi^2 + \Lambda^2}} \tag{3.63}$$

When $\eta^2 \ll 1$, one has simply

$$\eta \approx \frac{\Lambda}{2\,\pi} \tag{3.64}$$

Let us search again for a more convenient expression for the velocity $\dot{x}(t)$ than relation (3.54). By differentiating (3.57) we get

$$\dot{x} = -\,X\, e^{-\lambda t}(\lambda \cos(\omega_1 t-\varphi) + \omega_1 \sin(\omega_1 t-\varphi)) \tag{3.65}$$

This expression can be put into the form

$$\dot{x} = -\,X \sqrt{\lambda^2 + \omega_1^2}\; e^{-\lambda t} \sin(\omega_1 t - \varphi + \alpha)$$

in which

$$tg\alpha = \frac{\lambda}{\omega_1}$$

But $\lambda = \eta\,\omega_0$ and $\omega_1 = \omega_0\sqrt{1 - \eta^2}$, and so the velocity takes the simple form

$$\dot{x} = -\,\omega_0\,X\,e^{-\lambda t}\sin(\omega_1 t - \varphi + \alpha) \tag{3.66}$$

with

$$tg\alpha = \frac{\eta}{\sqrt{1 - \eta^2}} \tag{3.67}$$

If the damping factor is small, the term $\eta^2$ can be neglected compared to unity and consequently

$$\alpha \approx tg\,\alpha \approx \eta \qquad (\eta \ll 1) \tag{3.68}$$

Relation (3.66) means then that the velocity $\dot{x}$ is practically 90° out of phase to the displacement $x$, as is the case for a conservative oscillator (see figure 3.2, page 11).

The zeros of $\dot{x}(t)$ correspond to the extremes of $x(t)$ (alternatively a maximum and a minimum). Consequently, two successive maxima - minima of $x(t)$ are separated equally by the period $T_1$.

## 3.5 Energy of the dissipative oscillator

The total energy $H$ of the oscillator is the sum of the potential energy

$$V = \frac{1}{2}\,k\,x^2$$

and the kinetic energy.

$$T = \frac{1}{2}\,m\,\dot{x}^2$$

This is no longer a constant as for a conservative oscillator but a decreasing function because of the power dissipated by the damper. It gives, using (3.57) and (3.66)

$$H = \frac{1}{2} k X^2 e^{-2\lambda t} \cos^2(\omega_1 t - \phi) + \frac{1}{2} m \omega_0^2 X^2 e^{-2\lambda t} \sin^2(\omega_1 t - \phi + \alpha)$$

By definition $\omega_0^2 = k/m \Rightarrow m\omega_0^2 = k$ and the preceding expression can be written

$$H = \frac{1}{2} k X^2 e^{-2\lambda t}\left(\cos^2(\omega_1 t - \phi) + \sin^2(\omega_1 t - \phi + \alpha)\right)$$

or again

$$H = \frac{1}{2} k X^2 e^{-2\lambda t}\left(1 + \sin\alpha \cdot \sin 2(\omega_1 t - \phi + \frac{\alpha}{2})\right)$$

The angle $\alpha$ is given by (3.67)

$$\mathrm{tg}\ \alpha = \frac{\eta}{\sqrt{1 - \eta^2}} \quad \Rightarrow \quad \sin\alpha = \eta$$

whence finally

$$H = \frac{1}{2} k X^2 e^{-2\lambda t}\left(1 + \eta \sin 2(\omega_1 t - \beta)\right) \tag{3.69}$$

with

$$\beta = \phi - \frac{\alpha}{2} \quad \Rightarrow \quad \mathrm{tg}\ \beta = \frac{\eta\omega_0 X_0 + V_0(1 + \sqrt{1 - \eta^2})}{\eta V_0 + \omega_0 X_0(1 + \sqrt{1 - \eta^2})} \tag{3.70}$$

In this way, the total energy oscillates at the angular frequency $2\omega_1$ about a decreasing mean value

$$\bar{H} = \frac{1}{2} k X^2 e^{-2\lambda t} = \frac{1}{2} k (X e^{-\lambda t})^2 \tag{3.71}$$

equal to the potential energy corresponding to the envelope of the displacement (figure 3.12).

$$H = \bar{H}\ (1 + \eta \sin 2(\omega_1 t - \beta))$$

The first intersection between $H(t)$ and $\bar{H}(t)$ occurs at time $t'$ such that

$$t' = \frac{\beta}{\omega_1} \tag{3.72}$$

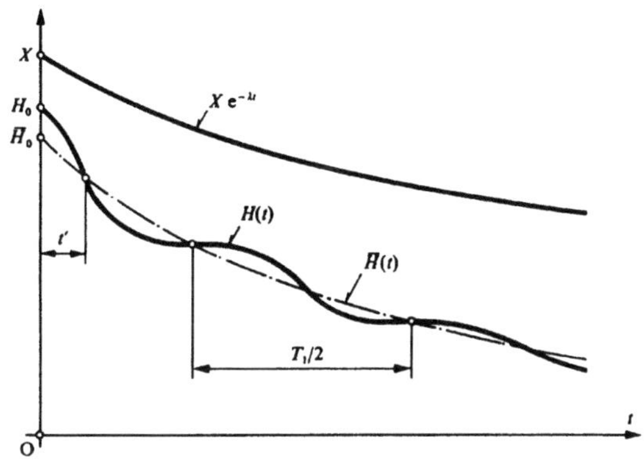


Fig. 3.12 Energy of a damped elementary oscillator

The energy lost by the oscillator can be calcuted as the difference between the initial energy

$$H_0 = \frac{1}{2}\,k\,X^2(1 + \eta\cos 2\beta) = \frac{1}{2}\,k\,X_0^2 + \frac{1}{2}\,m\,V_0^2 \tag{3.73}$$

and the energy $H$ at the time $t$

$$\Delta H_0^t = H_0 - H(t) \tag{3.74}$$

The power dissipated in the damper is equal to the product of the dissipative force and the velocity

$$p(t) = c\,\dot{x}\cdot\dot{x} = c\,\dot{x}^2 = c\,\omega_0^2\,X^2\,e^{-2\lambda t}\sin^2(\omega_1 t - \phi + \alpha) \tag{3.75}$$

The integral of this power between 0 and $t$ determines the energy lost

$$\Delta H_0^t = c\int_0^t \dot{x}^2 \, dt \tag{3.76}$$

When the damping factor is small ($\eta \ll 1$), one can interchange the energies $H$ and $\bar{H}$ without appreciable error. The energy lost then takes the simple form

$$\Delta H_0^t = \bar{H}_0 - \bar{H}(t) = \frac{1}{2} k X^2 (1 - e^{-2\lambda t}) \tag{3.77}$$

Likewise, the energy lost in an interval of time $\Delta t = t_2 - t_1$ has the value

$$\Delta H_{t_1}^{t_2} = \bar{H}(t_1) - \bar{H}(t_2) = \frac{1}{2} k X^2 (e^{-2\lambda t_1} - e^{-2\lambda t_2})$$

$$= \bar{H}(t_1)(1 - e^{-2\lambda\Delta t}) \tag{3.78}$$

## 3.6 Phase plane graph

In the preceding paragraph, we have calculated the energy of the oscillator as the sum of the potential and kinetic energies. It is a decreasing quantity. By designating the velocity by $v$, it has the value

$$\frac{1}{2} m v^2 + \frac{1}{2} kx^2 = H \tag{3.79}$$

or again, after division by $k/2$

$$\frac{v^2}{\omega_0^2} + x^2 = \frac{2H}{k} \tag{3.80}$$

The right-hand side of (3.80) has the same physical dimension as $x^2$ that is to say the square of a length

$$R^2 = \frac{2H}{k} \tag{3.81}$$

When the damping is zero, this length is a constant $R_0$. The equation (3.80) then becomes

$$\frac{v^2}{\omega_0^2} + x^2 = R_0^2 \tag{3.82}$$

It represents a circle of radius $R_0$ in the plane $x, v/\omega_0$, called the phase plane (figure 3.13). The circle is traced out in the sense of the hands of a watch since the displacement increases if the velocity is positive.

With non-zero damping, the radius $R = R(t)$ diminishes with time and the circle is transformed into a spiral which tends towards the origin.

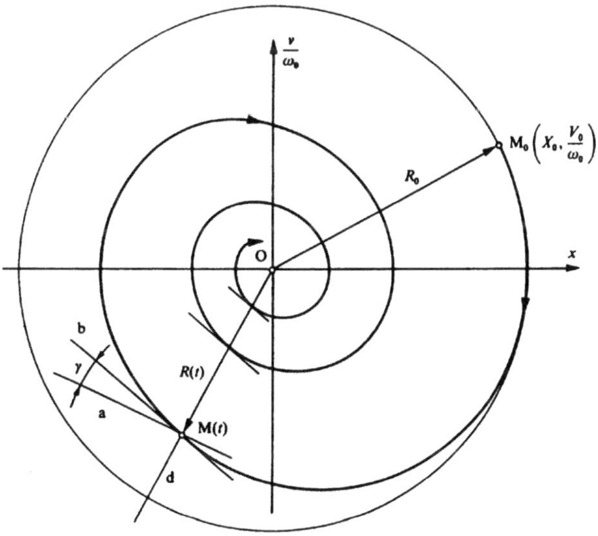


Fig. 3.13 Spiral of the phase plane (damping factor $\eta = 0.12$ )

The figure (3.13) shows the point $M(t)$ (given by the coordinates $x, v/\omega_0$), the radius vector $d$, its normal $a$, and the tangent to the spiral $b$. The angle $\gamma$ between $a$ and $b$ varies periodically but always keeps the same sign since $R(t)$ is always decreasing, except for $v = 0$. In effect, when the velocity is zero, the power dissipated is also zero and the energy $H$ is constant. Also, for all the points $M(t)$ situated on the axis $Ox$, the tangents to the spiral are vertical. (On the contrary, the tangents to $H(t)$ are horizontal at the corresponding points, see figure 3.12.)

The radius vector $d$ is an isocline; which means that tangents to the points of intersection between this radius vector and the spiral are all parallel. In order to demonstrate this, let us go back to the differential equation (3.32) by replacing $\dot{x}$ by $v$, and $\lambda$ by its value $\eta\omega_0$

$$\frac{dv}{dt} + 2\,\eta\,\omega_0 v + \omega_0^2\,x = 0 \tag{3.83}$$

By adopting the notation

$$y = \frac{v}{\omega_0} \quad \Rightarrow \quad \frac{dy}{dt} = \frac{1}{\omega_0}\frac{dv}{dt} \tag{3.84}$$

equation (3.83) can be put in the form

$$\frac{dy}{dt} = -\,(2\,\eta\,\omega_0 y + \omega_0\,x) \tag{3.85}$$

Let us calculate next the slope of the tangent to the spiral

$$\frac{dy}{dx} = \frac{dy}{dt}\frac{dt}{dx} = \frac{dy}{dt}\frac{1}{v} \tag{3.86}$$

which gives, taking account of (3.84) and (3.85)

$$\frac{dy}{dx} = -\,(2\,\eta + \frac{x}{y}) \tag{3.87}$$

The radius vector d passing through the origin, the ratio x/y is a constant which we call A . Consequently

$$\frac{dy}{dx} = -(2\eta + A) = B \tag{3.88}$$

Thus, the slope of the tangent is also a constant, which demonstrates that the radius vector d is really an isocline. This is the case for the horizontal axis in particular as we have previously asserted on the basis of physical reasoning.

Although analogues, figures 3.10 and 3.13 present several differences. In particular, in the first, the vertical projection of the vector $Xe^{-\lambda t}$ is not exactly proportional to the velocity whereas in the second the angle $M_0OM(t)$ is not exactly equal to $(\omega_1 t - \varphi)$ .

## 3.7 Examples of dissipative oscillators

### 3.7.1 Suspension element for a vehicle

A suspension system, intended for a road vehicle, is subject to laboratory tests which give the following results :

- the weight of a mass of 350 kg, equal to about one quarter of that of the vehicle, causes a static displacement $\delta = 0.28$ m (figure 3.14),
- The oscillations about the equilibrium position have a frequency $f_1 = 0.835$ Hz.

From these measurements, calculate the characteristics of the suspension element.

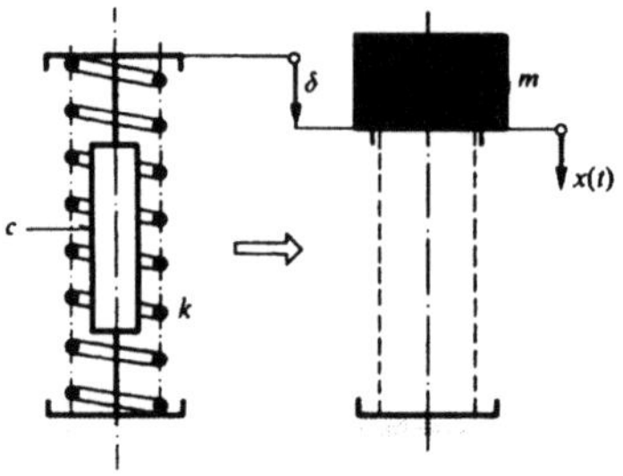


Fig. 3.14 Suspension element for a vehicle

One first calculates the stiffness of the spring

$$k\delta = mg \quad \Rightarrow \quad k = \frac{mg}{\delta} = \frac{350 \cdot 9.81}{0.28} = 12{,}260 \text{ N/m}$$

then the natural angular frequency of the undamped oscillator

$$\omega_0^2 = \frac{k}{m} = \frac{12{,}260}{350} = 35.04 \text{ rad}^2/\text{s}^2 \quad \Rightarrow \quad \omega_0 = 5.919 \text{ rad/s}$$

The natural angular frequency with damping has the value

$$\omega_1 = 2\pi\, f_1 = 2\pi \cdot 0.835 = 5.246 \text{ rad/s} \Rightarrow \quad \omega_1^2 = 27.52 \text{ rad}^2/\text{s}^2$$

One seeks the damping factor by means of relation (3.52)

$$\omega_1 = \omega_0 \sqrt{1-\eta^2} \quad \Rightarrow \quad \eta = \sqrt{1 - \frac{\omega_1^2}{\omega_0^2}} = \left(1 - \frac{27.52}{35.04}\right)^{1/2} = 0.463$$

The damping constant is deduced from the relation (2.4)

$$\eta = \frac{c}{2\, m\, \omega_0} \quad \Rightarrow \quad c = 2\, \eta\, m\, \omega_0 = 2 \cdot 0.463 \cdot 350 \cdot 5.919 = 1{,}920 \text{ kg/s}$$

In summary :

$k$ = 12,260 N/m

$c$ = 1,920 kg/s

$f_1$ = 0.835 Hz

$\eta$ = 0.463

} when the suspended mass is equal to 350 kg

### 3.7.2 Damping of a polymer bar

In order to measure the internal damping coefficient of a polymer bar, one records the oscillations in the free state of the system consisting of the bar and a mass m fixed at its end (figure 3.15). One notes that the amplitude of the sixth oscillation is equal to 30 % of that of the first. Using the following values :

- length $L = 1.5$ m
- modulus of elasticity $E = 2.2\ 10^9$ N/m²
- cross-section $A = 5$ cm² $= 0.5\ 10^{-3}$ m²
- mass $m = 100$ kg

determine the resistance c of the equivalent elementary oscillator (one neglects the mass of the bar), and then the internal damping constant τ of the polymer.

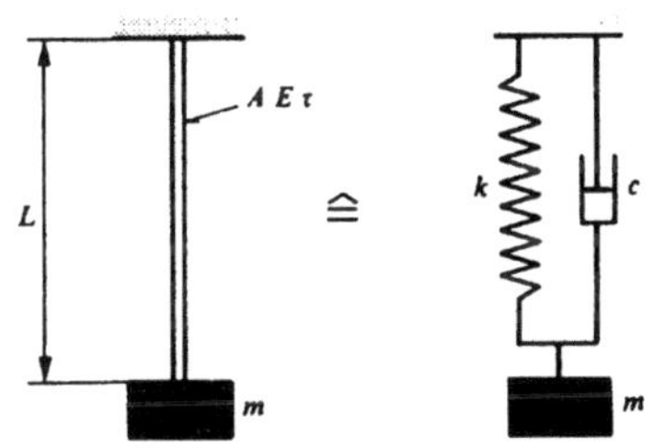


Fig. 3.15 Oscillator consisting of a polymer bar and a rigid mass

To neglect the mass of the bar compared to m is to go back to consider the system as an elementary oscillator whose stiffness and natural angular frequency are respectively

$$k = \frac{E\,A}{L} = \frac{2.2\cdot 10^9 \times 5\cdot 10^{-4}}{1.5} = 0.733\cdot 10^6 \text{ N/m}$$

$$\omega_0 = \sqrt{\frac{k}{m}} = \left(\frac{0.733\cdot 10^6}{100}\right)^{1/2} = 85.6 \text{ rad/s}$$

Between the first and the sixth oscillation, 5 periods $T_1$ have passed. The logarithmic decrement therefore has the value, according to equation (3.61),

$$\Lambda = \frac{1}{5} \ln \frac{1.00}{0.30} = 0.24079$$

The damping factor is deduced from (3.63)

$$\eta = \frac{\Lambda}{\sqrt{4\pi^2 + \Lambda^2}} = 0.03830 = 3.83 \%$$

By using the approximate relation (3.64), one obtains the very close value $\eta = 0.03832$ . The resistance of the equivalent oscillator can then be calculated by means of (2.4)

$$c = 2 \eta m \omega_0 = 2 \times 0.03829 \times 100 \times 85.6 = 656 \text{ kg/s}$$

To determine the internal damping constant of the material, it is sufficient to write down the equality between the viscous frictional forces in the bar and those in the equivalent oscillator

$$A \tau \dot{\varepsilon} = c \dot{x} \tag{3.89}$$

In this expression, $\dot{\varepsilon}$ is the derivative of the relative deformation (strain)

$$\dot{\varepsilon} = \frac{d\varepsilon}{dt} = \frac{d}{dt}\frac{x}{L} = \frac{1}{L}\dot{x} \tag{3.90}$$

It gives therefore

$$\tau = \frac{c L}{A} = \frac{656 \times 1.5}{5 \cdot 10^{-4}} = 1.97 \cdot 10^6 \text{ kg/ms} \tag{3.91}$$

Comments

- The example considered appears in the field of **rheological models** .

- By adopting the approximate relation (3.64), that is to say $\eta = \Lambda / 2\pi$ , it is easy to show that the coefficient $\tau$ can be calculated directly from the expression

$$\tau = \frac{\Lambda}{\pi} \sqrt{\frac{E\,L\,m}{A}} \tag{3.92}$$

### 3.7.3 Oscillator with dry friction

Calculate the number of half-periods made by the oscillator shown in figure 3.16, the mass being acted on by a dry friction with coefficient $\mu$ , the initial conditions being $x(0) = X_0$ and $\dot{x}(0) = 0$.

An oscillator subject to a dry frictional force, or Coulomb friction, is not a linear system. Nevertheless, the motion can be described by a succession of linear steps provided that one makes the assumption, which we are going to do, that the frictional force has a constant absolute value (even when the velocity is zero) and that it is in the direction opposed to the velocity. In these conditions, Newton's law applied to the mass gives

$$m\,\ddot{x} = -\,k\,x - (\mathrm{sgn}\,\dot{x})\,\mu\,mg \tag{3.93}$$

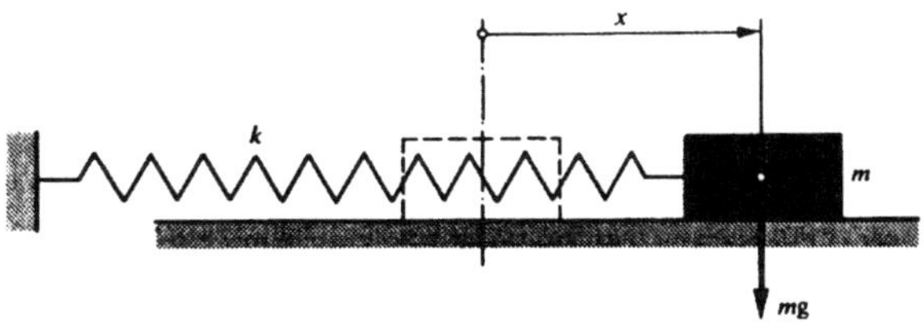


Fig. 3.16 Oscillator with dry friction

In this relation $(\text{sgn}\ \dot{x})$ is the sign of the velocity. The differential equation of motion is thus

$$m\ddot{x} + k\,x = -(\text{sgn}\ \dot{x})\ \mu\ mg \tag{3.94}$$

The general solution is the sum of the solution without the right-hand side, that is to say relation (3.3) and of a specific solution $x_p$ which one can choose constant

$$x_p = -(\text{sgn}\ \dot{x})\ \frac{\mu\ mg}{k} = -(\text{sgn}\ \dot{x})\ \frac{\mu\ g}{\omega_0^2} \tag{3.95}$$

The motion is therefore governed by the equation

$$x = X\cos(\omega_0 t - \varphi) - (\text{sgn}\ \dot{x})\ X_\ell \tag{3.96}$$

where $X_\ell = \frac{\mu\ mg}{k}$ represents displacement corresponding to the maximum dry frictional force. At displacements greater than $X_\ell$, the elastic restoring force is greater than the maximum dry frictional force.

$$\dot{x} = -\omega_0\ X\sin(\omega_0 t - \varphi) \tag{3.97}$$

With the initial conditions $x(0) = X_0$ and $\dot{x}(0) = 0$, the constants have the values, from (3.96) and (3.97)

$$X = X_0 + (\text{sgn}\ \dot{x})\ X_\ell \quad \text{and} \quad \varphi = 0 \tag{3.98}$$

As the system is in the free state, it is easy to see that, at the initial time $t = 0$, $(\text{sgn}\ \dot{x}) = -(\text{sgn}\ X_0)$. The motion is then written

$$x = (X_0 + (\text{sgn}\ \dot{x})\ X_\ell)\cos\omega_0 t - (\text{sgn}\ \dot{x})\ X_\ell \tag{3.99}$$

It is made up of harmonic half-periods, with the origin shifted by the quantity $X_\ell$ to the side of the positive $x$'s when the velocity is negative and vice versa.

By setting $u = 2t/T_0$, the index of the half-period $n$ is the integer whole of $u + 1$, which is to say

$$n = [u + 1] \tag{3.100}$$

Also with $X_0$ positive, the motion can be written

$$x = (X_0 - (2n-1)\ X_\ell)\cos \pi u - (-1)^n X_\ell \tag{3.101}$$

The extremes of decreasing motion are aligned on the straight lines of equation

$$x = \pm\ (X_0 - 2\ u\ X_\ell) \tag{3.102}$$

The movement of the mass stops at the end of the first half-period where

$$|x| \leqslant |X_\ell| \tag{3.103}$$

which gives, from (3.102)

$$X_0 - 2\ u\ X_\ell \leqslant X_\ell$$

or again

$$u \geqslant \frac{1}{2}\left(\frac{X_0}{X_\ell} - 1\right) \tag{3.104}$$

In this way the index $n_0$ of the last half-period made is obtained from (3.100)

$$n_0 = \left[\frac{1}{2}\left(\frac{X_0}{X_\ell} + 1\right)\right] \tag{3.105}$$

<u>Numerical example</u>

$m = 2$ kg  $k = 7{,}200$ N/m  $\mu = 0.32$

$X_0 = 1$ cm  $g = 9.81$ m/s$^2$

The limiting displacement $X_\ell$ then has the value

$$X_\ell = \mu \frac{mg}{k} = \frac{0.32 \times 2 \times 9.81}{7{,}200} = 0.872 \cdot 10^{-3} \text{ m}$$

The number of half-periods made is given by (3.105)

$$n_0 = \left[\frac{1}{2}\left(\frac{10}{0.872} + 1\right)\right] = [6.234] = 6$$

The mass stops therefore after six half-oscillations at the position $X_6$ which one can calculate from (3.101)

$$X_6 = (10 - 11 \times 0.872) \cos 6\pi - (-1)^6 \times 0.872$$

$$X_6 = - 0.464 \text{ mm} \qquad \text{and} \qquad |X_6| < X_\ell$$

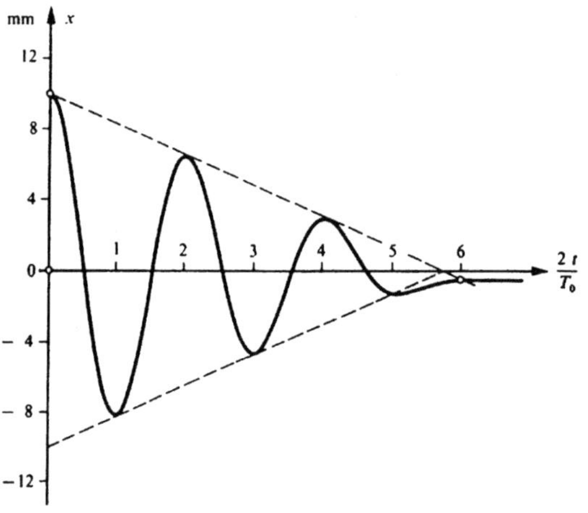


Fig. 3.17 Motion of the oscillator subject to dry friction, shown in figure 3.16

# CHAPTER 4

# HARMONIC STEADY STATE

Let us recall that the **forced state** is the behaviour of the oscillator when acted upon by an external force. The study of the forced state, in the general sense, is the subject of chapter 6. We will first consider a very important special case, that of the **harmonic steady state,** caused by a pure harmonic external force, after the disappearance of the transitory terms. It will then be possible to tackle, by means of the Fourier series, the **periodic steady state** due to any form of periodic force (chapter 5).

## 4.1 AMPLITUDE AND PHASE AS A FUNCTION OF FREQUENCY

Let us use again equation (2.1) with $f(t) = F \cos \omega t$

$$m \ddot{x} + c \dot{x} + k x = F \cos \omega t \tag{4.1}$$

and let us seek a steady state solution, therefore without transitory terms, of the form

$$x = A \cos \omega t + B \sin \omega t \tag{4.2}$$

By introducing $x$ and its derivatives into the equation, one obtains, by identification of the terms in $\cos \omega t$ and $\sin \omega t$

$$\begin{cases} A(k - \omega^2 m) + B \omega c = F \\ - A \omega c + B(k - \omega^2 m) = 0 \end{cases}$$

One easily extracts the constants $A$ and $B$ from it

$$\begin{cases} A = F \dfrac{k - \omega^2 m}{(k - \omega^2 m)^2 + \omega^2 c^2} \\ \\ B = F \dfrac{\omega c}{(k - \omega^2 m)^2 + \omega^2 c^2} \end{cases} \tag{4.3}$$

By combining the two harmonic functions (see fig. 3.1), the solution becomes

$$x = X \cos(\omega t - \varphi) \tag{4.4}$$

with $X = \sqrt{A^2 + B^2}$ and $tg\,\varphi = \dfrac{B}{A}$, which gives :

$$X = \frac{F}{\sqrt{(k - \omega^2 m)^2 + \omega^2 c^2}} \tag{4.5}$$

$$tg\,\varphi = \frac{\omega c}{k - \omega^2 m} \tag{4.6}$$

It is convenient, in order to study how $X$ and $\varphi$ vary as a function of $\omega$, to use the quantities introduced above :

$\omega_0^2 = \dfrac{k}{m}$ natural angular frequency of a conservative oscillator

$\eta = \dfrac{c}{2\, m\, \omega_0}$ damping factor

and, in addition, to define, the quantities :

$\beta = \dfrac{\omega}{\omega_0}$ **relative angular frequency** of an external force (4.7)

$X_s = \dfrac{F}{k}$ **static displacement** due to a constant force F (4.8)

$\mu = \dfrac{X}{X_s}$ **dynamic amplification factor** (4.9)

The relation (4.5) can be written

$$X = \frac{F/k}{\sqrt{(1 - \omega^2 \frac{m}{k})^2 + \frac{\omega^2 c^2}{k^2}}} = \frac{X_s}{\sqrt{(1 - (\frac{\omega}{\omega_0})^2)^2 + 4\,\eta^2 (\frac{\omega}{\omega_0})^2}}$$

By dividing by $X_s$ , one obtains the dynamic amplification factor which, taking into account (4.7), thus has the value

$$\mu = \frac{1}{\sqrt{(1 - \beta^2)^2 + 4\,\eta^2\,\beta^2}} \tag{4.10}$$

This is a dimensionless quantity which is equal, as we shall see in section 4.3, to the modulus of the **complex frequency response**. It is shown in figure 4.1 as a function of the relative angular frequency $\beta$ , with the damping factor $\eta$ as a parameter.

All the curves pass through the common point, $\beta = \omega = 0$, $\mu = 1$. This corresponds to the fact that a zero frequency force is a static force and that then, by definition, $X = X_s \Rightarrow \mu = 1$. They then pass through a maximum (except if $\eta \geqslant \sqrt{2}/2 = 0.707$) and then tend towards zero as the angular frequency goes towards infinity; the system stays immobile if one excites it infinitely quickly.

The oscillator is at an **amplitude resonance** when $X$ is maximum. One determines the corresponding angular frequency $\omega_2$ by finding the value which minimizes the denominator of equation (4.10) (and therefore its square).

$$\frac{d}{d\beta}\left((1 - \beta^2)^2 + 4\,\eta^2\,\beta^2\right) = 0 \qquad \Rightarrow \qquad \beta(-1 + \beta^2 + 2\,\eta^2) = 0$$

The solution $\beta = 0 \Rightarrow \omega = 0$ corresponds to the common point, already mentioned, which is a maximum of $\mu$ for $\eta \geqslant \sqrt{2}/2$ . The non-zero solution gives the angular frequency sought

$$\beta_2 = \sqrt{1 - 2\,\eta^2}\,,$$

and consequently

$$\omega_2 = \omega_0\sqrt{1 - 2\,\eta^2} \tag{4.11}$$

By introducing $\beta_2$ into (4.10), one obtains the maximum dynamic amplification factor

$$\mu_{max} = \frac{1}{2\,\eta\sqrt{1 - \eta^2}} \tag{4.12}$$

If $\beta = 1$, the same relation gives

$$\mu_0 = \frac{1}{2\,\eta} \tag{4.13}$$

When the damping is weak ($\eta \ll 1$), $\mu_0$ and $\mu_{max}$ have practically the same value, as one sees in figure 4.1.

It is useful to note that in the study of electrical oscillators, in particular in the theory of filters, one often uses the concept of **quality factor,** which is defined thus

$$Q = \frac{1}{2\,c/c'}$$

$c$ and $c'$ being respectively the resistance and the critical resistance. According to (2.4), one has

$$c = 2\,\eta\,m\,\omega_0 \quad \text{and} \quad c' = 2\,m\,\omega_0 \quad \Rightarrow \quad Q = \frac{1}{2\,\eta} = \mu_0$$

Thus, the quality factor is equal to the dynamic amplification factor for $\omega = \omega_0$.

Let us return now to the phase shift of the displacement on the external force (4.6)

$$\text{tg}\,\varphi = \frac{\omega\,c}{k - \omega^2\,m} = \frac{\omega\,\frac{c}{k}}{1 - \omega^2\,\frac{m}{k}} = \frac{2\,\eta\,\frac{\omega}{\omega_0}}{1 - (\frac{\omega}{\omega_0})^2}$$

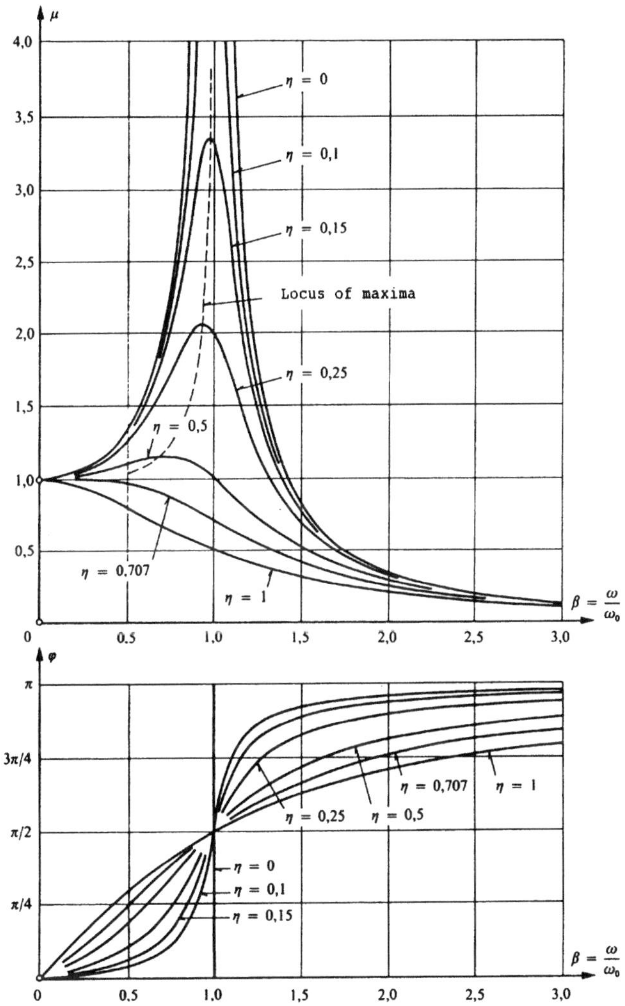


Fig. 4.1 Dynamic amplification factor $\mu$ and phase shift $\varphi$ as a function of the relative angular frequency $\beta$, with the damping factor $\eta$ as a parameter

By introducing the relative angular frequency $\beta = \frac{\omega}{\omega_0}$ , it gives

$$tg\ \varphi = \frac{2\ \eta\ \beta}{1 - \beta^2} \tag{4.14}$$

Whatever the value of the damping factor $\eta$ (figure 4.1), the external force and its displacement are

- in phase ($\varphi = 0$) if the angular frequency tends to zero,
- in quadrature ($\varphi = \frac{\pi}{2}$) if $\beta = 1 \quad \Rightarrow \quad \omega = \omega_0$
- in phase opposition ($\varphi = \pi$) if the angular frequency tends to infinity.

One says that the oscillator is at **phase resonance** when $\varphi = \frac{\pi}{2}$, thus for $\omega = \omega_0$.

## 4.2 GRAPH OF ROTATING VECTORS

It is interesting to find again the amplitude and the phase in the harmonic steady state by means of a graph of rotating vectors. Let us return then to (4.1)

$$m\ \ddot{x} + c\ \dot{x} + k\ x = F \cos \omega t$$

This relation expresses that the sum of the forces acting on the mass is zero. In the steady state $x = X \cos(\omega t - \varphi)$ and these forces have the value :

f(t) external force, in advance of the displacement $x$ by the phase $\varphi$

$$f(t) = F \cos \omega t$$

$f_m(t)$ force of inertia, in phase opposition to the displacement

$$f_m(t) = m\ \ddot{x} = -\ \omega^2\ m\ X \cos(\omega t - \varphi) = \omega^2\ m\ X \cos(\omega t - \varphi + \pi)$$

$f_c(t)$ viscous frictional force, in advance of the displacement by the phase of $\pi/2$ (in quadrature)

$$f_c(t) = c\,\dot{x} = -\,\omega\, c\, X \sin(\omega t - \varphi) = \omega\, c\, X \cos(\omega t - \varphi + \pi/2)$$

$f_k(t)$ elastic returning force, in phase with the displacement

$$f_k(t) = k\, x = k\, X \cos(\omega t - \varphi)$$

These forces are equal to the projections onto an axis of vectors rotating at the same angular velocity $\omega$ (fig. 4.2).

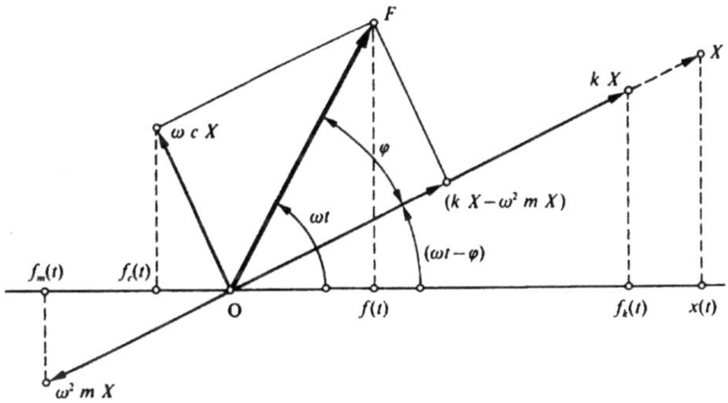


Fig. 4.2 Harmonic steady state; graph of rotating vectors.

The unknowns of the problem are the amplitude $X$ and the phase $\varphi$ which the graph enables one to calculate easily

$$(k\,X - \omega^2\, m\, X)^2 + (\omega\, c\, X)^2 = F^2 \quad \Rightarrow \quad X = \frac{F}{\sqrt{(k - \omega^2\, m)^2 + \omega^2\, c^2}}$$

$$\mathrm{tg}\ \varphi = \frac{\omega\, c}{k - \omega^2\, m}$$

One clearly finds again the results (4.5) and (4.6).

## 4.3 Use of complex numbers. Frequency Response

A more efficient method of calculation than the above, although based upon the same basic idea, consists of replacing the vectors by complex numbers. This method, developed originally by electrical engineers, will often be used in the following text. For the moment, with the objective of presenting the principle of the process and of defining some important concepts, we will calculate again the amplitude and phase of the displacement of an elementary oscillator subject to an harmonic force.

Let us go back to equation (4.1) for the motion

$$m\ddot{x} + c\dot{x} + kx = F\cos\omega t$$

and let us search for a steady state solution of the form

$$x = X\cos(\omega t - \varphi)$$

in which X and $\varphi$ are the unknowns to be calculated.

The external force $f = F\cos\omega t$ can be considered as the real part (figure 4.3) of the **complex force** which is defined as follows

$$\underline{f} = F\,e^{j\omega t} = F(\cos\omega t + j\sin\omega t) \qquad (4.15)$$

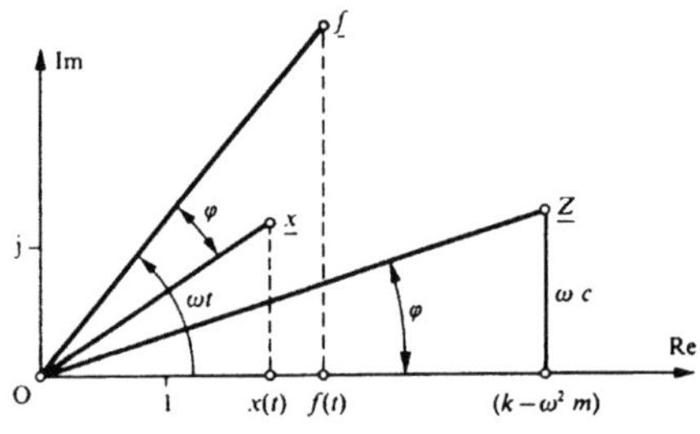


Fig. 4.3 Graph in the complex plane

and one has easily

$$f = \mathrm{Re}(\underline{f})$$

The displacement x(t) is then the real part of the **complex displacement**

$$\underline{x} = X\, e^{j(\omega t - \varphi)} \qquad x = \mathrm{Re}(\underline{x}) \tag{4.16}$$

Equation (4.1) represents the real part of the complex equation

$$m\,\ddot{\underline{x}} + c\,\dot{\underline{x}} + k\,\underline{x} = \underline{f} \tag{4.17}$$

One first calculates the derivatives of x

$$\dot{\underline{x}} = j\,\omega\,X\,e^{j(\omega t - \varphi)} = j\,\omega\,\underline{x}$$

$$\ddot{\underline{x}} = -\,\omega^2\,X\,e^{j(\omega t - \varphi)} = -\,\omega^2\,\underline{x}$$

whence by introduction into the previous equation

$$((k - \omega^2 m) + j\,\omega\,c)\,\underline{x} = \underline{f} \tag{4.18}$$

The quantity

$$\underline{Z} = (k - \omega^2 m) + j\,\omega\,c = \sqrt{(k - \omega^2 m)^2 + \omega^2 c^2}\; e^{j\varphi} \tag{4.19}$$

is called the **complex impedance**. Its inverse is the **complex admittance**

$$\underline{Y} = \frac{1}{\underline{Z}} = \frac{1}{(k - \omega^2 m) + j\,\omega\,c} = \frac{e^{-j\varphi}}{\sqrt{(k - \omega^2 m)^2 + \omega^2 c^2}} \tag{4.20}$$

The angle $\varphi$ is given by its tangent

$$\mathrm{tg}\,\varphi = \frac{\omega\,c}{k - \omega^2 m}$$

Equation (4.17) finally reduces to

$$\underline{Z}\,\underline{x} = \underline{f}$$

which has the solution

$$\underline{x} = \frac{\underline{f}}{\underline{Z}} = \underline{Y}\,\underline{f} \tag{4.21}$$

$$\underline{x} = \frac{F\,e^{j\omega t}}{(k - \omega^2 m) + j\,\omega\,c} = \frac{F\,e^{j(\omega t - \varphi)}}{\sqrt{(k - \omega^2 m)^2 + \omega^2 c^2}} \tag{4.22}$$

By comparing (4.16) and (4.22), one obtains the previous results, explicitly

$$X = \frac{F}{\sqrt{(k - \omega^2 m)^2 + \omega^2 c^2}} \qquad \mathrm{tg}\,\varphi = \frac{\omega\,c}{k - \omega^2 m}$$

Relation (4.20) shows that the physical dimensions of the stiffness and of the complex admittance are the inverse of each other (the brackets [] mean "dimension of") :

$$[k] = \frac{\text{Newton}}{\text{metre}} = \frac{N}{m} \qquad [\,\underline{Y}\,] = \frac{m}{N}$$

So the product

$$\underline{H} = k\,\underline{Y} \tag{4.23}$$

is without dimension. It is the **complex frequency response**, whose modulus $\mu$ has already been introduced in paragraph 4.1 under the name of **dynamic amplification factor**. In fact, by using (4.20) and the usual notation, one obtains

$$\underline{H}(\omega) = \frac{k}{(k-\omega^2 m) + j\,\omega\,c} = \frac{1}{(1-\omega^2 \frac{m}{k}) + j\,\frac{\omega\,c}{k}} = \frac{1}{(1 - (\frac{\omega}{\omega_0})^2) + 2\,j\eta\,\frac{\omega}{\omega_0}} \tag{4.24}$$

then, by introducing the relative angular frequency,

$$\underline{H}(\beta) = \frac{1}{(1 - \beta^2) + 2\, j\, \eta\, \beta} \tag{4.25}$$

One can equally write

$$\begin{cases} \underline{H} = \mu\, e^{-j\varphi} = \dfrac{e^{-j\varphi}}{\sqrt{(1 - \beta^2)^2 + 4\, \eta^2\, \beta^2}} \\ \operatorname{tg} \varphi = \dfrac{2\, \eta\, \beta}{1 - \beta^2} \end{cases} \tag{4.26}$$

It is convenient to use the complex frequency response and the static displacement in (4.21)

$$\underline{x} = (k\, \underline{Y})\, \frac{1}{k}\, \underline{f} = (k\, \underline{Y})\, \frac{F}{k}\, e^{j\omega t}$$

$$\underline{x} = \underline{H}\, X_s\, e^{j\omega t} \tag{4.27}$$

Let us finally note that the quantity $X_s\, e^{j\omega t}$ represents the complex form $\underline{x}_e$ of the **elastic displacement** defined by relation (2.6). That being the case, (4.27) can be written

$$\underline{x} = \underline{H}\, \underline{x}_e \tag{4.28}$$

## 4.4 Power consumed in the steady state

The instantaneous power supplied to the system is equal to the product of the external force times the velocity :

$$p(t) = f(t)\, \dot{x}(t) = -\, F \cos \omega t \cdot \omega\, X \sin(\omega t - \varphi)$$

which gives by expansion

$$p(t) = -\, \omega\, X\, F \cos \varphi \cos \omega t \sin \omega t + \omega\, X\, F \sin \varphi \cos^2 \omega t$$

The first term is the **reactive power** which corresponds to a zero energy loss per period. The second term is the **active power**, effectively consumed by the damper and which leads to a loss of energy per period amounting to

$$H_d = \omega X F \sin\varphi \int_0^T \cos^2\omega t \, dt = \pi X F \sin\varphi$$

One can eliminate $X$ and $\varphi$ using the relations (4.5) and (4.6)

$$X = \frac{F}{\sqrt{(k - \omega^2 m)^2 + \omega^2 c^2}}$$

$$\operatorname{tg}\varphi = \frac{\omega c}{k - \omega^2 m} \quad \Rightarrow \quad \sin\varphi = \frac{\operatorname{tg}\varphi}{\sqrt{1 + \operatorname{tg}^2\varphi}} = \frac{\omega c}{\sqrt{(k - \omega^2 m)^2 + \omega^2 c^2}}$$

One obtains in this way

$$H_d = \frac{\pi \omega c F^2}{(k - \omega^2 m)^2 + \omega^2 c^2} \tag{4.29}$$

The **mean power consumed** - or **effective power** - can be deduced immediately

$$\bar{p} = \frac{H_d}{T} = \frac{H_d \omega}{2\pi} = \frac{\omega^2 c F^2}{2((k - \omega^2 m)^2 + \omega^2 c^2)} \tag{4.30}$$

In order to study how $\bar{p}$ varies as a function of the driving angular frequency and of the damping, it is convenient to take its ratio to the power $\bar{p}_0$ consumed by the oscillator when $\omega = \omega_0$ and $\eta = 1$ (critical damping). Let us recall that

$$\omega_0^2 = \frac{k}{m}$$

$$\eta = \frac{c}{2 m \omega_0}$$

so for $\eta = 1$, $c = 2 m \omega_0$ and

$$\bar{p}_0 = \frac{F^2}{4 m \omega_0} \tag{4.31}$$

One can then define the relative power

$$\varepsilon = \frac{\bar{p}}{\bar{p}_0} = \frac{2 m \omega_0 \omega^2 c}{(k - \omega^2 m)^2 + \omega^2 c^2}$$

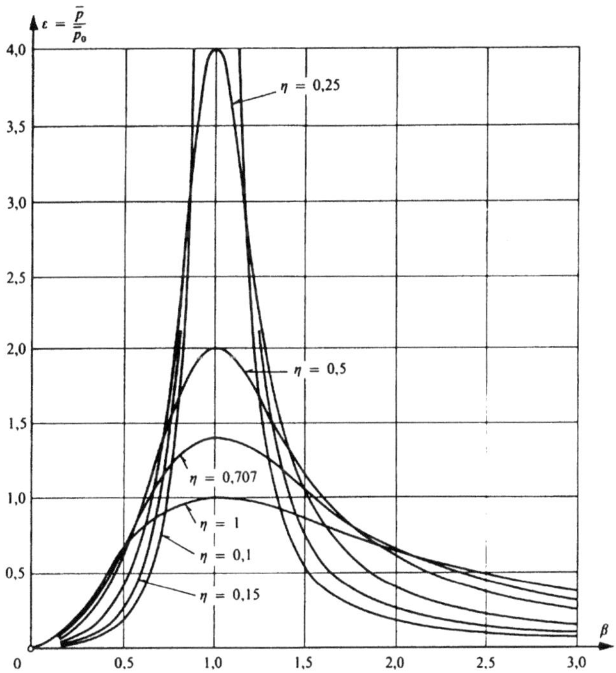


Fig. 4.4 Relative power $\varepsilon$ consumed in the steady state as a function of the relative angular frequency $\beta$, with the damping factor $\eta$ as a parameter

By introducing the factors $\eta$ and $\beta$ , it becomes

$$\varepsilon = \frac{4\ \eta\ \beta^2}{(1 - \beta^2)^2 + 4\ \eta^2\ \beta^2} \qquad (4.32)$$

The relative power has a maximum value for $\omega = \omega_0$ that is to say (fig. 4.4),

$$\varepsilon_{max} = \frac{1}{\eta} \qquad (4.33)$$

In this way the **power resonance,** like the **phase resonance,** occurs at the angular frequency $\omega_0$, whatever the damping factor.

## 4.5 Natural and resonant angular frequencies

The linear elementary oscillator has four notable angular frequencies of which three are already known to us ($\omega_0$, $\omega_1$, $\omega_2$). Before summarizing the situation, which is the purpose of this section, it is necessary to determine the angular frequencies for the **velocity resonance** and the acceleration **resonance** in the steady state.

One knows, from the results of paragraph 4.1, that the steady state displacement is given by the expression

$$x = X\cos(\omega t - \varphi) \quad , \quad \text{with}$$

$$X = \mu\ X_s = \frac{X_s}{\sqrt{(1 - \beta^2)^2 + 4\ \eta^2\ \beta^2}}$$

The velocity is thus

$$\dot{x} = -\ \omega\ X\sin(\omega t - \varphi) = -\ V\sin(\omega t - \varphi)$$

Let us write its amplitude V

$$V = \omega\ X = \omega_0\ X_s\ \frac{\beta}{\sqrt{(1 - \beta^2)^2 + 4\ \eta^2\ \beta^2}} \qquad (4.34)$$

then let us seek the angular frequency for which the oscillator is at **velocity resonance,** that is to say for which $V$ is maximum

$$\frac{\partial V}{\partial \beta} = 0 \qquad \Rightarrow \qquad (1 - \beta^2)\cdot(1 + \beta^2) = 0$$

The relative angular frequency $\beta$ being by nature real and positive, the only useful root of the equation above is $\beta = 1$. Thus, the velocity resonance occurs at $\omega = \omega_0$.

By proceeding in the same way for the acceleration, this method gives successively

$$\ddot{x} = -\omega^2 X \cos(\omega t - \varphi) = -A \cos(\omega t - \varphi)$$

$$A = \omega^2 X = \omega_0^2 X_S \frac{\beta^2}{\sqrt{(1 - \beta^2)^2 + 4 \eta^2 \beta^2}} \qquad (4.35)$$

$$\frac{\partial A}{\partial \beta} = 0 \qquad \Rightarrow \qquad \beta^2(1 - 2\eta^2) - 1 = 0$$

Let us call $\beta_3$ the solution to this equation

$$\beta_3 = \frac{1}{\sqrt{1 - 2\eta^2}}$$

The **acceleration resonance** would appear therefore for an angular frequency $\omega_3$ greater than $\omega_0$ (4.36)

$$\omega_3 = \frac{\omega_0}{\sqrt{1 - 2\eta^2}}$$

To summarize, a linear elementary oscillator has the following notable angular frequencies :

$\omega_0 = \sqrt{\frac{k}{m}}$ — natural angular frequency without damping, angular frequency at the phase resonance, at the velocity resonance and at the power resonance

$\omega_1 = \omega_0 \sqrt{1 - \eta^2}$ natural angular frequency with damping

$\omega_2 = \omega_0 \sqrt{1 - 2\eta^2}$ angular frequency at the amplitude resonance

$\omega_3 = \dfrac{\omega_0}{\sqrt{1 - 2\eta^2}}$ angular frequency at the acceleration resonance

They are classified in the following order

$$\omega_2 < \omega_1 < \omega_0 < \omega_3 \tag{4.37}$$

It is thus useful, when one speaks of the resonant angular frequency of an oscillator, to specify which one is referred too. If one does not do this, it is in principle to $\omega_0$ to which one refers.

## 4.6 The Nyquist graph

The curve of the figure 4.5 shows the complex frequency response $\underline{H}$ as a function of the relative angular frequency $\beta$, for a constant value of the damping factor $\eta$. One such curve is a Nyquist graph. According to the relations (4.25) and (4.26), $\underline{H}$ is given by the expression

$$\underline{H} = \frac{1}{(1 - \beta^2) + 2j\eta\beta} = \mu\, e^{-j\varphi}$$

with

$$\mu = \frac{1}{\sqrt{(1 - \beta^2)^2 + 4\eta^2\beta^2}} \qquad \mathrm{tg}\varphi = \frac{2\eta\beta}{1 - \beta^2}$$

The curve $\underline{H}(\beta)$ approximates to a circle, especially when the damping factor is small (figure 4.7). It is shown plotted for increasing values of $\beta$ from $\beta = 0$ (point s) to $\beta = \infty$ (point 0) in figure 4.5.

Let us designate the real and imaginary parts of $\underline{H}$ by a and b

$$\underline{H} = a + j\,b \tag{4.38}$$

They have respectively the values

$$\begin{cases} a = \dfrac{1 - \beta^2}{(1 - \beta^2)^2 + 4\eta^2\beta^2} \\[2ex] b = \dfrac{-2\,\eta\,\beta}{(1 - \beta^2)^2 + 4\eta^2\beta^2} \end{cases} \tag{4.39}$$

Since the imaginary part is always negative, only the lower half of figure 4.5 is accessible, with the exception, however, of the half-circle of diameter OS, as we shall see below. The notable angular frequencies, listed at the end of the previous paragraph, correspond to the points $\underline{H}_0$, $\underline{H}_1$, $\underline{H}_2$ and $\underline{H}_3$, in the order shown in the figure as a result of the inequalities (4.37).

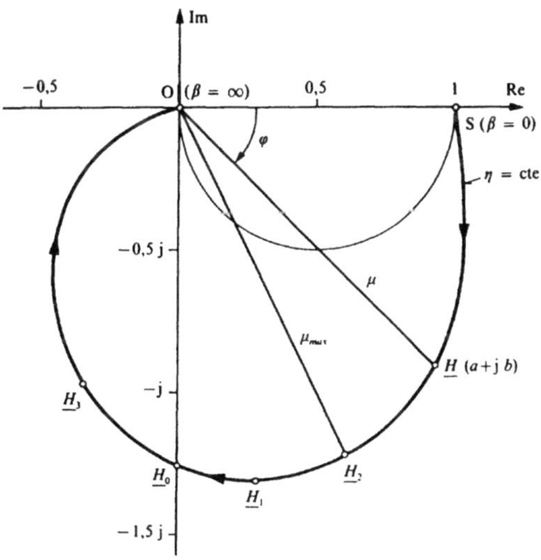


Fig. 4.5 Nyquist graph of $\underline{H}(\beta)$ (with $\eta = 0.4$ )

For the point $\underline{H}_0$ (corresponding to the phase, velocity and power resonances $\beta_0 = 1$ ), the real part is zero when the imaginary part is

$$a_0 = \frac{-1}{2\eta} \; (= -\mu_0) \qquad (4.40)$$

For the point $\underline{H}_2$ (amplitude resonance $\beta_2 = \sqrt{1 - 2\eta^2}$ ), the modulus is equal to the maximum (4.12) and one obtains

$$\begin{cases} a_2 = \dfrac{1}{2(1 - \eta^2)} \\[2ex] b_2 = \dfrac{-1\sqrt{1 - 2\eta^2}}{2\eta(1 - \eta^2)} \end{cases} \qquad (4.41)$$

When the damping factor tends to zero, $a_2$ tends towards 1/2. Thus, the locus of the points $\underline{H}_2$ , shown in figure 4.7, has a vertical asymptote. Finally, one can prove that the imaginary part $b_1$ of $\underline{H}_1$ is very close to a maximum.

Let us now choose the damping factor as a parameter, the relative angular frequency being assumed to be constant. One can show that the point $\underline{H}$ describes a semi-circle in the complex plane (figure 4.6), corresponding to the equation

$$\begin{cases} (a - R)^2 + b^2 = R^2 \\[2ex] R = \dfrac{1}{2(1 - \beta^2)} \end{cases} \qquad (4.42)$$

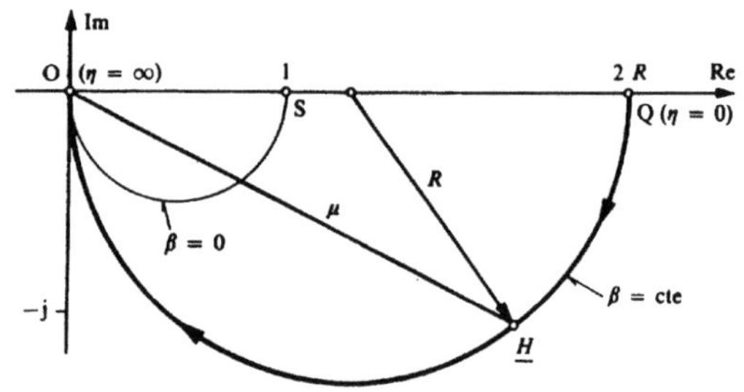


Fig. 4.6 Locus of $\underline{H}$ as a function of $\eta$ ($\beta = 0.8$)

When the damping factor increases, the semi-circle is traced out from the point Q ($\eta = 0$) to the point O ($\eta = \infty$) . For $\beta = 0$ , the diameter is equal to unity and the interior of the half-circle is inaccessible.

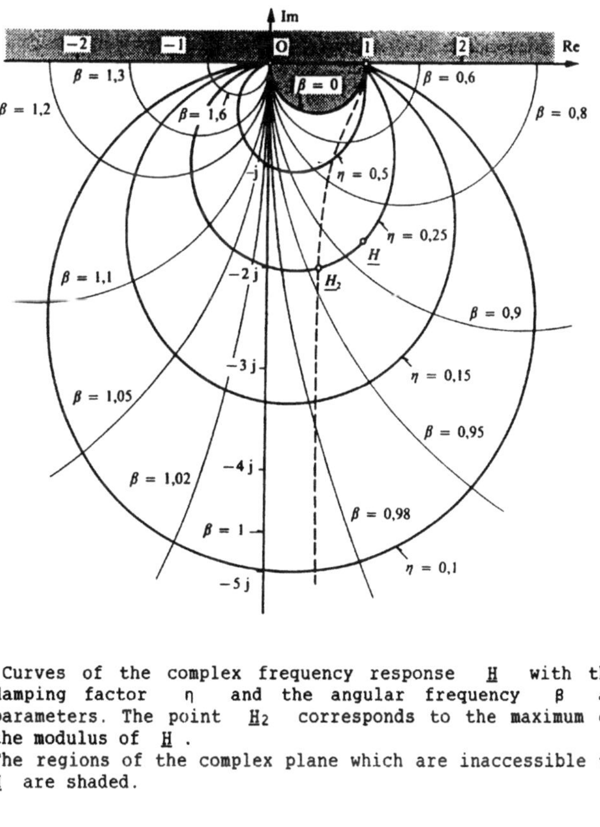


Fig. 4.7 Curves of the complex frequency response $\underline{H}$ with the damping factor $\eta$ and the angular frequency $\beta$ as parameters. The point $\underline{H}_2$ corresponds to the maximum of the modulus of $\underline{H}$ .
The regions of the complex plane which are inaccessible to $\underline{H}$ are shaded.

## 4.7 Examples of harmonic steady states

### 4.7.1 Vibrator for fatigue tests

Figure 4.8 shows schematically a vibrator subjecting a test piece (E) to a compression fatigue test. The exciting force is supplied by two unbalanced weights rotating at the angular velocity $\omega$ . The force $f'(t)$ on the test piece, of amplitude $F'$ , is transmitted by means of a plate attached with nuts and bolts.

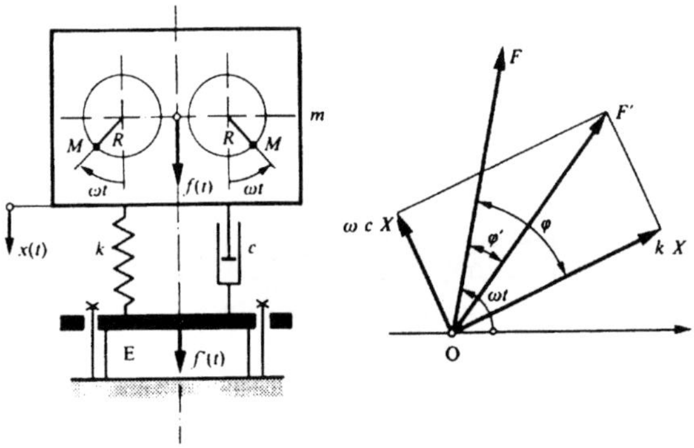


Fig. 4.8 Vibrator for fatigue tests

At what speed must the unbalanced weights rotate so that F' = 3200N, assuming that the test piece is rigid in comparison with the spring?

R = 5 cm   M = 2 kg   m = 80 kg (including the unbalanced weights)

k = 700,000 N/m   $\eta$ = 0.2

The excitation force due to the unbalanced weights is

$$f(t) = 2\ R\ M\ \omega^2 \cdot \cos \omega t = F \cos \omega t \tag{4.43}$$

In the steady state, which is the only one considered here, it causes a displacement of the mass

$$x = X \cos(\omega t - \varphi)$$

whose amplitude is given by (4.5)

$$X = \frac{F}{\sqrt{(k - \omega^2\ m)^2 + \omega^2\ c^2}} \tag{4.44}$$

The force transmitted to the test tube is thus

$$f'(t) = k\ x + c\ \dot{x} = k\ X \cos(\omega t - \varphi) - \omega\ c\ X \sin(\omega t - \varphi)$$
$$= k\ X \cos(\omega t - \varphi) + \omega\ c\ X \cos(\omega t - \varphi + \frac{\pi}{2})$$

One can put it in the form (figure 4.8)

$$f'(t) = F' \cos(\omega t - \varphi')$$

Only the amplitude F' interests us

$$F' = X \sqrt{k^2 + \omega^2\ c^2} \tag{4.45}$$

It gives, by using (4.43) and (4.44)

$$F' = 2\ R\ M\ \omega^2 \sqrt{\frac{k^2 + \omega^2\ c^2}{(k - \omega^2\ m)^2 + \omega^2\ c^2}}$$

With the usual notation

$$\omega_0^2 = \frac{k}{m} \qquad \eta = \frac{c}{2\ m\ \omega_0} \qquad \beta = \frac{\omega}{\omega_0}$$

the preceding relation takes the form

$$F' = (2\ R\ M\ \omega_0^2)\ \beta^2 \sqrt{\frac{1 + 4\ \eta^2\ \beta^2}{(1 - \beta^2)^2 + 4\ \eta^2\ \beta^2}} = (2\ R\ M\ \omega_0^2)\ \alpha(\beta) \tag{4.46}$$

In the particular case envisaged $\eta = 0.2$ the dimensionless function $\alpha(\beta)$ becomes

$$\alpha(\beta) = \beta^2 \sqrt{\frac{1 + 0.16\ \beta^2}{(1 - \beta^2)^2 + 0.16\ \beta^2}}$$

This is shown in figure 4.9 (if one studies $\alpha(\beta)$ with $\eta$ as a parameter, one obtains a family of curves all passing through the common point $\beta = \sqrt{2}$, $\alpha = 2$ ).

One has successively

$$\omega_0^2 = \frac{k}{m} = \frac{700{,}000}{80} = 8{,}750\ s^{-2} \quad \Rightarrow \quad \omega_0 = 93.5\ s^{-1}$$

$$2\ R\ M\ \omega_0^2 = 2 \times 0.05 \times 2 \times 8{,}750 = 1{,}750\ N$$

$$(4.46) \rightarrow \alpha(\beta) = \frac{F'}{2\ R\ M\ \omega_0^2} = \frac{3{,}200}{1{,}750} = 1.829$$

The figure shows that there are 3 values of $\beta$ for which $\alpha = 1.829$. One calculates them by solving the equation

$$1.829^2 = \beta^4\ \frac{1 + 0.16\ \beta^2}{(1 - \beta^2)^2 + 0.16\ \beta^2}$$

and one finds

$$\beta_1 = 0.862 \qquad \beta_2 = 1.567 \qquad \beta_3 = 3.384$$

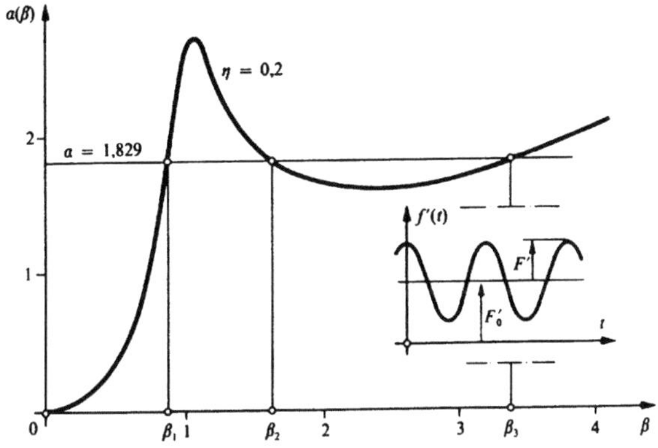


Fig. 4.9 Relative force $\alpha$ as a function of the relative angular frequency $\beta$

The corresponding rates of rotation, in revolutions per minute, have the values

$$n_i = \beta_i \frac{\omega_0}{2\pi} 60 = \beta_i \frac{93.5}{2\pi} 60$$

$n_1 = 770$ rpm $\qquad n_2 = 1{,}400$ rpm $\qquad n_3 = 3{,}020$ rpm

Let us note again that the force B of traction of the nuts and bolts, regulated by elastic washers, must be sufficient to ensure permanent contact between the mounting plate and the test tube.

$$F_0' = mg + B \quad > \quad F' = 3{,}200 \text{ N}$$

In the case $F_0' = 1.5\ F'$ , the load on the test tube has the shape shown in figure 4.9. Such a load is called **alternating compression.**

Comments

- In practice, it is in general easier to adjust the alternating force by changing the radius R of the unbalanced weights.

· Test machines in which the force is produced by hydraulic jacks naturally present considerable advantages over the mechanism examined in this example. They are also much more expensive.

4.7.2 Measurement of damping

Experimental determination of the the damping factor $\eta$ of an elementary oscillator by means of a test in the harmonic steady state (figure 4.10).

After having measured the amplitude $X(\omega)$ of the displacement as a function of the exciting angular frequency, one draws a horizontal straight line at the ordinate $1/\sqrt{2}\, X_{max}$ . This straight line determines the points M' and M" corresponding to the angular frequencies $\omega'$ and $\omega''$ (band width at half-power). Let us establish the approximate relation.

$$\eta = \frac{\omega'' - \omega'}{\omega'' + \omega'} .$$

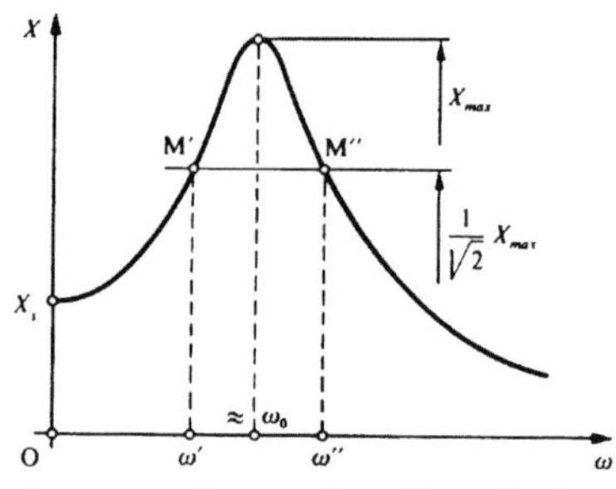


Fig. 4.10 Amplitude measured as a function of the angular frequency

Let us return to the relations (4.10) and (4.12)

$$\mu = \frac{X}{X_0} = \frac{1}{\sqrt{(1 - \beta^2)^2 + 4\,\eta^2\,\beta^2}} \qquad \mu_{max} = \frac{X_{max}}{X_s} = \frac{1}{2\,\eta\,\sqrt{1 - \eta^2}}$$

The relative angular frequencies $\beta' = \frac{\omega'}{\omega_0}$ and $\beta'' = \frac{\omega''}{\omega_0}$ are thus fixed by the condition

$$\frac{1}{\sqrt{2}} \cdot \frac{1}{2\,\eta\,\sqrt{1-\eta^2}} = \frac{1}{\sqrt{(1-\beta^2)^2 + 4\,\eta^2\,\beta^2}}$$

which gives

$$\beta^2 = (1 - 2\,\eta^2) \pm 2\,\eta\,\sqrt{1-\eta^2} \tag{4.47}$$

By first assuming that the damping factor is small $(\eta^2 \ll 1)$, one can simplify the above equation

$$\beta^2 = 1 \pm 2\,\eta$$

and as a result

$$\begin{aligned} \beta'^2 &= 1 - 2\,\eta \\ \beta''^2 &= 1 + 2\,\eta \end{aligned} \quad \Rightarrow \quad \eta = \frac{1}{4}\,(\beta''^2 - \beta'^2) = \frac{1}{4}\,(\beta'' + \beta')\,(\beta'' - \beta')$$

Moreover $\quad \eta \ll 1 \quad \Rightarrow \quad \omega_0 = \frac{1}{2}\,(\omega'' + \omega') \quad \Rightarrow \quad \beta'' + \beta' = 2$

$$\eta = \frac{1}{2}\,(\beta'' - \beta') \tag{4.48}$$

By expressing this result as a function of the angular frequencies measured $\omega'$ and $\omega''$, one obtains the result sought

$$\eta = \frac{1}{2}\,\frac{\omega'' - \omega'}{\omega_0} = \frac{\omega'' - \omega'}{\omega'' + \omega'} \tag{4.49}$$

By introducing the ratio

$$\alpha = \frac{\omega'}{\omega''} \tag{4.50}$$

the preceding relation becomes

$$\eta = \frac{1 - \alpha}{1 + \alpha} \tag{4.51}$$

Exact solution

Let us write the relation (4.47) for the angular frequencies $\omega'$ and $\omega''$

$$\beta'^2 = \frac{\omega'^2}{\omega_0^2} = 1 - 2\,\eta^2 - 2\,\eta\,\sqrt{1 - \eta^2} \tag{4.52}$$

$$\beta''^2 = \frac{\omega''^2}{\omega_0^2} = 1 - 2\,\eta^2 + 2\,\eta\,\sqrt{1 - \eta^2} \tag{4.53}$$

Taking account of (4.50), it gives after division term by term and expansion

$$\eta^4 - \eta^2 + \frac{(1 - \alpha^2)^2}{8(1 + \alpha^4)} = 0$$

Once the acceptable solution of this equation is known, one determines $\omega_0$ by (4.52) or (4.53).

Numerical example

$$\omega' = 114\ s^{-1} \qquad \omega'' = 126\ s^{-1} \qquad \Rightarrow \qquad \alpha = \frac{114}{126} = 0.9048$$

Let us denote the approximate values by $\eta_a$ and $\omega_{0a}$

$$\eta_a = \frac{126 - 114}{126 + 114} = 0.05 = 5\ \%$$

$$\omega_{0a} = \frac{1}{2}\,(114 + 126) = 120\ s^{-1}$$

The exact values are respectively

$$\eta_e = 4.969\ \% \qquad \omega_{0e} = 120.45\ s^{-1}$$

The approximate method leads in this way to relative errors :

· for the damping factor

$$\varepsilon(\eta) = \frac{\eta_a - \eta_e}{\eta_e} = \frac{5 - 4.969}{4.969} = +\ 0.62\ \%\ \text{(over estimate)}\ :$$

· for the resonance frequency

$$\varepsilon(\omega_0) = \frac{\omega_{0a} - \omega_{0e}}{\omega_{0e}} = \frac{120 - 120.45}{120.45} = -\ 0.37\ \%\ \text{(under estimate)}$$

Comments

· The range of application of the approximate method is quite large. In effect :

$$\varepsilon(\eta) \leqslant +\ 5\ \% \quad \Rightarrow \begin{cases} \alpha \geqslant 0.75 & (\alpha = \frac{\omega'}{\omega''}) \\ \eta \leqslant 0.14 & \\ |\varepsilon(\omega_0)| \leqslant 2.9\ \% & \end{cases}$$

· In practice, measurement of the damping factor by means of a test in the free state (examples 3.7.1 and 3.7.2) is usually preferable to measurement in the steady state described above.

### 4.7.3 Vibrations of a machine shaft

The shaft of a machine supports a disk at its centre of mass $m_1$, having a mass unbalance of $m_2$ (figure 4.11). Determine the ranges of speeds of rotation for which the maximum bending stress, from vertical motion, stays below a value $\sigma_0$ specified, by taking into account the stress due to the weight of the flywheel, but neglecting the mass of the shaft. Let us use the following numerical values

| | | |
|---|---|---|
| $m_1$ = 200 kg | L = 0.8 m | E = $2.1 \cdot 10^{11}$ Pa |
| $m_2$ = 0.2 kg | R = 0.25 m | $\sigma_0$ = 100 MPa |
| g = 9.81 m/s$^2$ | r = 0.03 m | |

It is important to point out that the solution of this problem must not be considered as a method for fatigue rating of a machine shaft.

<u>Fig. 4.11</u>
Shaft of machine supporting an unbalanced disk

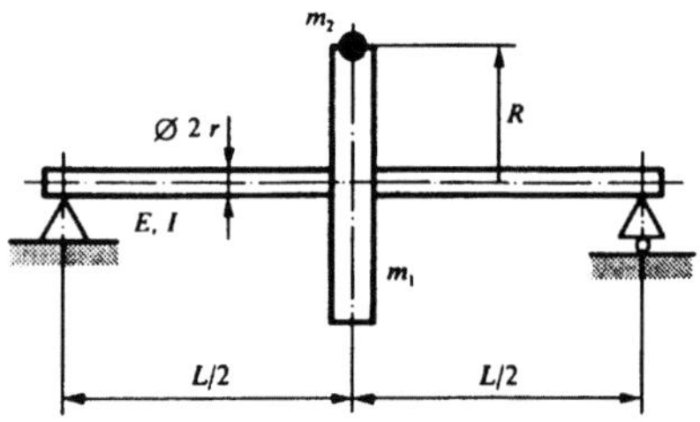


As the damping is assumed to be zero and $m_2 \ll m_1$ , the differential equation for vertical motion $x_2(t)$ of the disk can be written

$$m_1 x_2 + k x_2 = m_2 R \omega^2 \cos \omega t \tag{4.54}$$

The result (4.10) gives the amplitude $X_2$ of the steady state displacement

$$X_2 = \frac{m_2 R}{k} \omega^2 \frac{1}{|1 - \beta^2|} \tag{4.55}$$

$\beta = \omega/\omega_0$ being the relative angular frequency.

By substituting $\delta$ for the expression

$$\delta = \frac{m_2}{m_1} R$$

relation (4.55) takes the form

$$X_2 = \delta \frac{\beta^2}{|1 - \beta^2|} \tag{4.56}$$

The methodology of the strength of materials enables one to calculate the deflection and the maximum bending stress, at the centre of the beam due to the static weight of the disk

$$X_1 = \frac{m_1 \, g \, L^3}{48 \, E \, I} \qquad \sigma_1 = \frac{m_1 \, g \, r \, L}{4 \, I} \tag{4.57}$$

In these expressions, E and I are respectively the modulus of elasticity and the moment of inertia for the bending of the shaft. By eliminating the weight, it gives

$$X_1 = \frac{\sigma_1 \, L^2}{12 \, E \, r} \tag{4.58}$$

The maximum deflection at the centre of beam, corresponding to the stress $\sigma_0$ , is thus

$$X_0 = \frac{\sigma_0 \, L^2}{12 \, E \, r} \tag{4.59}$$

The sum of the stress due to the static weight and of that due to the unbalanced weight varies as the direction of the latter stress changes with the rotation. This sum has a maximum when the unbalanced weight acts downwards with the static weight. This leads to the condition :

$$X_1 + X_2 \leqslant X_0 \tag{4.60}$$

By introducing the results (4.56), (4.58) and (4.59) into relation (4.60), we get

$$\delta \frac{\beta^2}{|1 - \beta^2|} + \frac{\sigma_1 \, L^2}{12 \, E \, r} \leqslant \frac{\sigma_0 \, L^2}{12 \, E \, r}$$

that is to say

$$\frac{\beta^2}{|1 - \beta^2|} \leqslant (\sigma_0 - \sigma_1) \frac{L^2}{12 \, E \, r \, \delta} \tag{4.61}$$

The numerical values enable us to calculate

$$I = \frac{\pi \, r^4}{4} = 6.362 \cdot 10^{-7} \text{ m}^4 \qquad \delta = \frac{0.2}{200} \, 0.25 = 2.5 \cdot 10^{-4} \text{ m}$$

then successively

$$\sigma_1 = \frac{200 \times 9.81 \times 0.8 \times 3\cdot10^{-2}}{4 \times 6.362\cdot10^{-7}} = 1.850\cdot10^{7} \text{ Pa}$$

$$X_1 = \frac{200 \times 9.81 \times 0.8^3}{48 \times 2.1\cdot10^{11} \times 6.362\cdot10^{-7}} = 1.567\cdot10^{-4} \text{ m}$$

The second term of the condition (4.61) has the value

$$(10 - 1.85)\ 10^7 \frac{0.8^2}{12 \times 2.1\cdot10^{11} \times 3\cdot10^{-2} \times 2.5\cdot10^{-4}} = 2.760$$

which leads to the equation

$$\frac{\beta^2}{|1 - \beta^2|} \leq 2.76$$

from which one derives $\beta \leq 0.857$
$\beta \geq 1.252$

The natural angular frequency and the natural frequency of the system are by definition

$$\omega_0 = \sqrt{\frac{k}{m_1}} = \sqrt{\frac{g}{X_1}} = 250.2 \text{ s}^{-1}$$

$$f_0 = \frac{\omega_0}{2\pi} = 39.8 \text{ Hz}$$

The equivalent speed of rotation has the value

$$n_0 = 60 \cdot f_0 = 2{,}390 \text{ rpm}$$

The ranges of speeds allowed are finally : $n \leq 2048$ rpm and $n \geq 2992$ rpm.

As figure 4.12 shows, when $n$ becomes very large, the maximum displacement tends towards $\delta + X_1$, which corresponds to a bending stress of $\sigma = 48$ MPa.

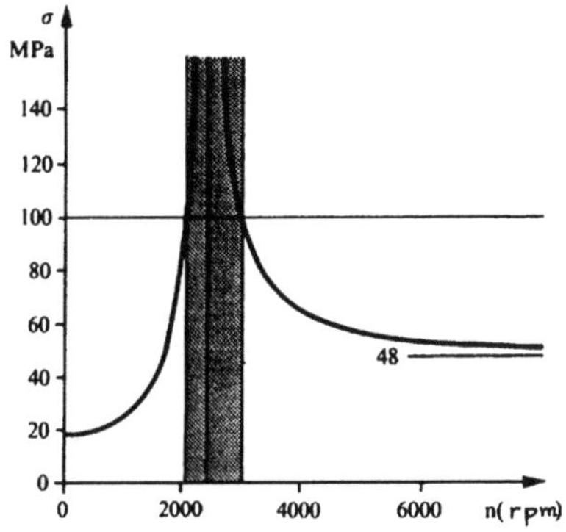


Figure 4.12 Bending stress as a function of the speed of the shaft shown in figure 4.11.

# CHAPTER 5 PERIODIC STEADY STATE

## 5.1 Fourier series · Excitation and response spectra

Harmonic analysis allows one to calculate the steady state of a dissipative oscillator excited by any periodic external force f(t). We mean, by the steady state, that state which maintains itself after the disappearance of transitory terms. . Let us recall that f(t) of period $T = \frac{2\pi}{\omega}$ (figure 5.1), can be decomposed into the Fourier series.

$$f(t) = \frac{1}{2} F_0 + \sum_n^{\infty} (A_n \cos n\omega t + B_n \sin n\omega t) \tag{5.1}$$

In this expression, the index n varies from one to infinity while the coefficients are given by the integrals

$$\begin{cases} F_0 = \frac{2}{T} \int_0^T f(t)\, dt \\ A_n = \frac{2}{T} \int_0^T f(t) \cos n\omega t\, dt \\ B_n = \frac{2}{T} \int_0^T f(t) \sin n\omega t\, dt \end{cases} \tag{5.2}$$

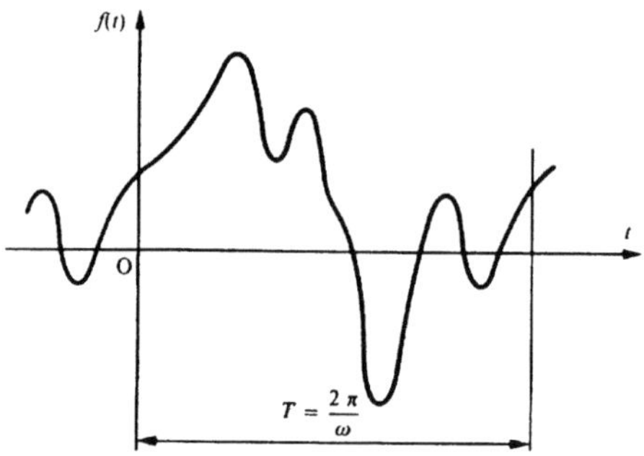


Fig. 5.1 Periodic external force

When the function $f(t)$ is even $(f(-t) = f(t))$, the constants $B_n$ are zero and the series consists only of cosine terms. Conversely, if $f(t)$ is odd $(f(-t) = -f(t))$, the constants $A_n$ are zero and the series consists only of sine terms.

By grouping the cosines and the sines for the same angular frequency, $f(t)$ becomes

$$f(t) = \frac{1}{2} F_0 + \sum_{n}^{\infty} F_n \cos(n\omega t - \psi_n) \tag{5.3}$$

with

$$\begin{cases} F_n = \sqrt{A_n^2 + B_n^2} \\ \operatorname{tg} \psi_n = \dfrac{B_n}{A_n} \end{cases} \tag{5.4}$$

The term $F_1 \cos(\omega t - \psi_1)$ is the **fundamental** of the external force; the terms of higher order $(n > 1)$ are called **harmonics**.

Let us return now to the equation for the oscillator

$$m \ddot{x} + c \dot{x} + k x = f(t)$$

It is a linear differential equation, which allows one to superpose the particular solutions corresponding to each term of (5.3). One will have then

$$x(t) = \frac{1/2 \, F_0}{k} + \sum_{n}^{\infty} X_n \cos(n\omega t - \psi_n - \varphi_n) \tag{5.5}$$

This result shows that $x(t)$ is a periodic function, with the same period as $f(t)$, and so justifies the expression **periodic steady state.**

The amplitude $X_n$ of the nth harmonic can be calculated by means of the relation (4.5)

$$X_n = \frac{F_n}{\sqrt{(k - n^2 \omega^2 m)^2 + n^2 \omega^2 c^2}} = \mu_n X_{sn} \tag{5.6}$$

By analogy with (4.8) and (4.10), the static displacement and the corresponding dynamic amplification factor have the values

$$X_{sn} = \frac{F_n}{k} \tag{5.7}$$

$$\mu_n = \frac{1}{(1 - n^2 \beta^2)^2 + 4 \eta^2 n^2 \beta^2} \tag{5.8}$$

As for the phase shift, it is given by (4.6) or (4.14)

$$\mathrm{tg}\, \varphi_n = \frac{n \omega c}{k - n^2 \omega^2 m} = \frac{2 \eta n \beta}{1 - n^2 \beta^2} \tag{5.9}$$

Figure 5.2 shows how the series of the amplitudes $F_n$ (spectrum of f(t)) are transformed into a series of amplitudes $X_n$ (spectrum of x(t) ) by means of the following two examples :

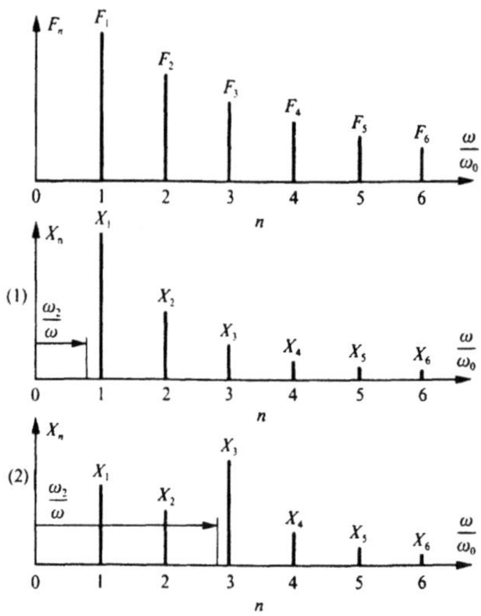


Fig. 5.2 Spectra of f(t) and x(t)
(1) $\omega_2/\omega < 1$ (2) $\omega_2/\omega \approx 3$

(1) The angular frequency at the amplitude resonance $\omega_2$ is less than that of the fundamental of $f(t)$ . So a decreasing series $F_n$ is transformed into a more strongly decreasing series $X_n$ .

(2) The angular frequency $\omega_2$ is close to $3\omega$ , for example. In this case the amplitude $X_3$ of the third harmonic of $x(t)$ can become greater than $X_1$ , even if $F_3 < F_1$.

The amplitudes of the harmonics of $x(t)$ decrease more rapidly than those of the harmonics of $f(t)$ , as soon as their angular frequencies $n\omega$ are greater than about one and a half times the angular frequency $\omega_2$ (see figure 4.1). The oscillator behaves like a high frequency filter. This effect would appear clearly by writing the relations (5.8) and (5.9) as follows

$$\mu_n = \frac{1}{n^2} \frac{1}{\sqrt{(\frac{1}{n^2} - \beta^2)^2 + \frac{4\, n^2\, \beta^2}{n^2}}}$$

$$\text{tg}\ \varphi_n = \frac{1}{n} \frac{2\, n\, \beta}{\frac{1}{n^2} - \beta^2}$$

For large values of $n$ , they become

$$\mu_n \approx \frac{1}{n^2\, \beta^2}$$

$$\text{tg}\ \varphi_n \approx - \frac{2\, n}{n\, \beta}$$

Thus, the amplitude $X_n$ decreases as $\frac{1}{n^2}$ and the phase shift $\varphi_n$ tends to $\pi$.

## 5.2 Complex form of the Fourier series

By using the Euler relations linking trigonometric functions to complex exponential functions

$$\cos n\omega t = \frac{e^{jn\omega t} + e^{-jn\omega t}}{2} \qquad \sin n\omega t = \frac{e^{jn\omega t} - e^{-jn\omega t}}{2j}$$

decomposition into the Fourier series takes the form

$$f(t) = \frac{1}{2} F_0 + \frac{1}{2} \sum_{n}^{\infty} (A_n(e^{jn\omega t} + e^{-jn\omega t}) - j\, B_n(e^{jn\omega t} - e^{-jn\omega t}))$$

or, by grouping the terms differently

$$f(t) = \frac{1}{2} F_0 + \frac{1}{2} \sum_{n}^{\infty} ((A_n - j\, B_n)\, e^{jn\omega t} + (A_n + j\, B_n)\, e^{-jn\omega t})$$

Let us introduce the notation

$$\begin{cases} C_0 = \frac{1}{2} F_0 \\ \underline{C}_n = \frac{1}{2} (A_n - j\, B_n) \\ \underline{C}_n^* = \underline{C}_{-n} = \frac{1}{2} (A_n + j\, B_n) \end{cases} \tag{5.10}$$

So the series becomes

$$f(t) = C_0 + \sum_{n}^{\infty} (\underline{C}_n\, e^{jn\omega t} + \underline{C}_n^*\, e^{-jn\omega t}) \tag{5.11}$$

or, by summing from $-\infty$ to $+\infty$

$$f(t) = \sum_{-\infty}^{\infty} \underline{C}_n\, e^{jn\omega t} \tag{5.12}$$

The complex coefficients $\underline{C}_n$ can be calculated from (5.2)

$$\underline{C}_n = \frac{1}{2} (A_n - j\, B_n) = \frac{1}{T} \int_0^T f(t)\, (\cos n\omega t - j \sin n\omega t)\, dt$$

$$\underline{C}_n = \frac{1}{T} \int_0^T f(t)\, e^{-jn\omega t}\, dt \qquad n = 0,1,2,\ldots \tag{5.13}$$

The preceding relations represent the complex - or exponential - form the Fourier series.

One can free oneself from the negative terms of the summation (5.12) by considering $f(t)$ as the real part of a complex force $\underline{f}(t)$ defined as follows

$$f(t) = \text{Re}\, \underline{f}(t) = \text{Re}(D_0 + \sum_n^{\infty} \underline{D}_n\, e^{jn\omega t}) \tag{5.14}$$

To find the values of the complex coefficients $\underline{D}_n$, it is sufficient to equate the previous relation to (5.11) which one can write as follows

$$f(t) = C_0 + \sum_n^{\infty} ((\text{Re}\, \underline{C}_n + j\, \text{Im}\, \underline{C}_n)\, e^{jn\omega t} + (\text{Re}\, \underline{C}_n - j\, \text{Im}\, \underline{C}_n)\, e^{-jn\omega t})$$

$$f(t) = C_0 + 2 \sum_n^{\infty} (\text{Re}\, \underline{C}_n \cos n\omega t - \text{Im}\, \underline{C}_n \sin n\omega t) \tag{5.15}$$

In the same way, (5.14) can be put in the form

$$f(t) = \text{Re}(D_0 + \sum_n^{\infty} (\underline{D}_n \cos n\omega t + j\, \underline{D}_n \sin n\omega t))$$

$$f(t) = D_0 + \sum_n^{\infty} (\text{Re}\, \underline{D}_n \cos n\omega t - \text{Im}\, \underline{D}_n \sin n\omega t) \tag{5.16}$$

The equality of (5.15) and (5.16) gives, taking account of (5.13)

$$D_0 = C_0$$
$$\underline{D}_n = 2\,\underline{C}_n = \frac{2}{T}\int_0^T f(t)\, e^{-jn\omega t}\, dt \tag{5.17}$$

In relation (5.14), the term $\underline{D}_n\, e^{jn\omega t}$ represents the nth harmonic of the exciting force. It is comparable to the complex force $F\, e^{j\omega t}$ introduced in section 4.3, apart from the difference that $\underline{D}_n$ is a **complexe amplitude** with a phase shift $\psi_n$.

$$\underline{D}_n = D_n\, e^{-j\psi_n} \qquad \Rightarrow \qquad \underline{D}_n\, e^{jn\omega t} = D_n\, e^{j(n\omega t - \psi_n)}$$

Comparison with (5.3) and (5.4) shows that

$$D_n = F_n = \sqrt{A_n^2 + B_n^2} \qquad\qquad \mathrm{tg}\, \psi_n = \frac{B_n}{A_n}$$

We have seen in section 4.3 that the response of an oscillator to an excitation $f(t) = \mathrm{Re}(F\, e^{j\omega t})$ is equal to the real part of the **complex displacement** (or **complex response**)

$$x(t) = \mathrm{Re}\ \underline{X}(t) \tag{5.18}$$

This complex displacement has the value

$$\underline{X}(t) = \underline{Y}\, F\, e^{j\omega t} = \underline{H}\, X_s\, e^{j\omega t}$$

Let us recall that in these expressions, $\underline{Y}$ , $\underline{H}$ and $X_s$ are respectively the complex admittance, the complex frequency response and the static displacement.

If the exciting force has harmonics, the linearity of the system allows one to superpose the displacements due to these harmonics. On the other hand, the relation (5.18) remains valid. Therefore we get

$$\underline{x}(t) = \frac{D_0}{k} + \sum_n^{\infty} \underline{x}_n(t) = \frac{D_0}{k} + \sum_n^{\infty} \underline{Y}_n \underline{D}_n e^{jn\omega t}$$

$$= \frac{D_0}{k} + \sum_n^{\infty} \underline{H}_n \frac{1}{k} \underline{D}_n e^{jn\omega t} \tag{5.19}$$

One defines **the static displacement** $X_{sn}$ and the **complex static displacement** $\underline{X}_{sn}$ caused by the nth harmonic of the force. The phase shift of this harmonic being $\psi_n$ , one has

$$\underline{X}_{sn} = \frac{1}{k} \underline{D}_n = \frac{1}{k} D_n e^{-j\psi_n} = X_{sn} e^{-j\psi_n} \tag{5.20}$$

The phase shift of the displacement, the complex admittance and the complex frequency response for the nth harmonic can be deduced from the relations (4.14), (4.20) and (4.25) by replacing $\omega$ by $n\omega$ and $\beta$ by $n\beta$ $(\beta = \frac{\omega}{\omega_0})$.

$$\text{tg } \varphi_n = \frac{2 \eta n \beta}{1 - n^2 \beta^2} \tag{5.21}$$

$$\underline{Y}_n = \frac{1}{(k - n^2 \omega^2 m) + j n \omega c} = \frac{e^{-j\varphi_n}}{\sqrt{(k - n^2 \omega^2 m)^2 + n^2 \omega^2 c^2}} \tag{5.22}$$

$$\underline{H}_n = \frac{1}{(1 - n^2 \beta^2) + 2 j \eta n \beta} \tag{5.23}$$

Let us finally show the agreement between the relations (5.5) and (5.6) on the one hand and (5.19) on the other. The product $\underline{Y}_n \underline{D}_n$ can be written

$$\underline{Y}_n \cdot \underline{D}_n = \frac{D_n}{\sqrt{(k - n^2 \omega^2 m)^2 + n^2 \omega^2 c^2}} e^{-j(\varphi_n + \psi_n)}$$

Since $D_n = F_n$ , one finds the amplitude $X_n$ of the nth harmonic of the displacement.

$$X_n = \frac{F_n}{\sqrt{(k - n^2 \omega^2 m)^2 + n^2 \omega^2 c^2}}$$

Equation (5.5) thus becomes with $D_0 = 1/2\ F_0$

$$\underline{x}(t) = \frac{1/2\ F_0}{k} + \sum_n^{\infty} X_n\ e^{j(n\omega t - \psi_n - \varphi_n)} \tag{5.24}$$

The real part of (5.24) is clearly identical to (5.5).

## 5.3 Examples of periodic steady states

### 5.3.1 Steady state beats

Study the displacement $x(t)$ produced by two harmonic forces with similar amplitudes and angular frequencies.

The two forces

$$f'(t) = F' \cos \omega' t \qquad \text{and} \qquad f''(t) = F'' \cos \omega'' t$$

cause steady state displacements $x'(t)$ and $x''(t)$ whose amplitudes and phases can be calculated by means of relations (4.5) and (4.6). The total displacement is therefore

$$x = X' \cos(\omega' t - \varphi') + X'' \cos(\omega'' t - \varphi'') \tag{5.25}$$

It is convenient to express the amplitudes $X'$ and $X''$ as a function of their sum and of the difference

$$x = \frac{1}{2}(X' + X'')\,(\cos(\omega' t - \varphi') + \cos(\omega'' t - \varphi''))$$

$$+ \frac{1}{2}(X' - X'')\,(\cos(\omega' t - \varphi') - \cos(\omega'' t - \varphi''))$$

Let us transform the sums of cosines into products

$$x = (X' + X'')\cos\left(\frac{\omega' + \omega''}{2}t - \frac{\varphi' + \varphi''}{2}\right) \cdot \cos\left(\frac{\omega' - \omega''}{2}t - \frac{\varphi' - \varphi''}{2}\right)$$

$$- (X' - X'')\sin\left(\frac{\omega' + \omega''}{2}t - \frac{\varphi' + \varphi''}{2}\right) \cdot \sin\left(\frac{\omega' - \omega''}{2}t - \frac{\varphi' - \varphi''}{2}\right) \tag{5.26}$$

In order to simplify the writing, we adopt the notation

$$\omega = \frac{1}{2}(\omega' + \omega'') \qquad \varphi = \frac{1}{2}(\varphi' + \varphi'')$$
$$\alpha = \frac{1}{2}(\omega' - \omega'') \qquad \psi = \frac{1}{2}(\varphi' - \varphi'') \tag{5.27}$$

Taking account of (5.27), the relation (5.28) becomes

$$x = (X' + X'')\cos(\omega t - \varphi)\cdot\cos(\alpha t - \psi)$$
$$- (X' - X'')\sin(\omega t - \varphi)\cdot\sin(\alpha t - \psi) \tag{5.28}$$

The displacement $x$ oscillates with an angular frequency $\omega$ inside an envelope which itself oscillates, at the lower angular frequency $\alpha$, between the extreme amplitudes $X'+X''$ and $X'-X''$. This behaviour, called **steady state beats,** is shown in figure 5.3 for the following two cases :

Case (a) $X' = 0.7$ $X'' = 0.3$
Case (b) $X' = X'' = 0.5$
} $\Rightarrow X' + X'' = 1.0$

Case (a) and (b) $\omega = 50$ rad/s $\alpha = 3.5$ rad/s
$\varphi = 0.4$ rad $\psi = 0.1$ rad

When the amplitudes $X'$ and $X''$ are equal, the displacement takes the simple form

$$x = 2X'\cos(\omega t - \varphi)\cdot\cos(\alpha t - \psi)$$

It becomes zero for each half-period $\tau/2 = \pi/\alpha$, with a velocity which is itself zero. The oscillator is then momentarily at rest without any energy at all.

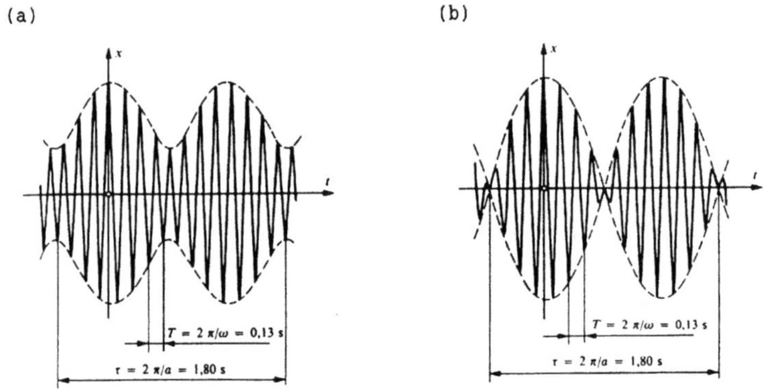


Fig. 5.3 Steady state beats
(a) X' = 0.7 X" = 0.3
(b) X' = X" = 0.5

5.3.2 Response to a periodic rectangular excitation

The external force f(t) shown in figure 5.4 acts on an elementary oscillator in the steady state. Find the spectrum of f(t) , then that of x(t) for the two following cases

(a) $\omega_0 = 0.8\ \omega$ $\eta = 0.05$
(b) $\omega_0 = 5.3\ \omega$ $\eta = 0.05$

In the particular case envisaged, the function f(t) is even and its mean value is zero. By referring to relations (5.2), one has therefore

$$F_0 = 0 \qquad B_n = 0 \qquad \Rightarrow \qquad A_n = F_n \qquad \text{and} \qquad \psi_n = 0$$

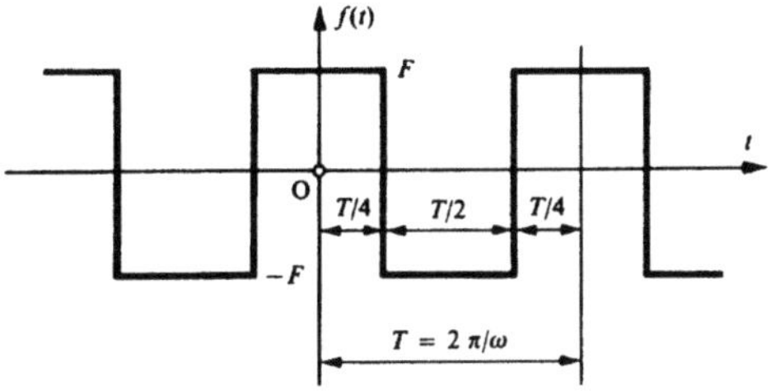


Fig. 5.4 Periodic rectangular external force

One then finds easily

$$A_n = F \frac{4}{n\pi} \sin \frac{n\pi}{2} \quad \Rightarrow \quad \begin{cases} A_n = \pm F \dfrac{4}{n\pi} & \text{if } n \text{ odd} \\ A_n = 0 & \text{if } n \text{ even} \end{cases} \tag{5.29}$$

The exciting force can be written

$$f(t) = \sum_{n}^{\infty} F_n \cos n\omega t \tag{5.30}$$

with

$$F_n = F \frac{4}{n\pi} \sin \frac{n\pi}{2} \tag{5.31}$$

The amplitudes of the harmonics of $f(t)$ are inversely proportional to their order. The ratio $F_n/F$ is given in the table of figure 5.5.

The response of the system is obtained by superposition of harmonic solutions, that is to say by (5.5),

$$x(t) = \sum_{n}^{\infty} X_n \cos(n\omega t - \varphi_n) \tag{5.32}$$

with, using (5.6), (5.7) and (5.8)

$$X_n = X_{sn}\,\mu_n = \frac{F_n}{k}\,\frac{1}{\sqrt{(1 - n^2\,\beta^2)^2 + 4\,\eta^2\,n^2\,\beta^2}} \tag{5.33}$$

By referring to the static displacement, which would cause a constant force of magnitude F, by $\delta = F/k$, the relative amplitude of motion of the nth harmonic becomes

$$\frac{X_n}{\delta} = \frac{4}{n\,\pi}\,\frac{1}{\sqrt{(1 - n^2\,\beta^2)^2 + 4\,\eta^2\,n^2\,\beta^2}} \tag{5.34}$$

Case (a)

$\omega_0 = 0.8\,\omega \quad \Rightarrow \quad \beta = \frac{\omega}{\omega_0} = 1.25 \quad \eta = 0.05$

$$\frac{X_n}{\delta} = \frac{4}{n\,\pi}\,((1 - 1.5625\,n^2)^2 + 1.563\cdot 10^{-2}\,n^2)^{-1/2}$$

Case (b)

$\omega_0 = 5.3\,\omega \quad \Rightarrow \quad \beta = 0.1887 \quad \eta = 0.05$

$$\frac{X_n}{\delta} = \frac{4}{n\,\pi}\,((1 - 3.56\cdot 10^{-2}\,n^2)^2 + 3.56\cdot 10^{-4}\,n^2)^{-1/2}$$

One can in this way establish the table shown in figure 5.5

| n | 1 | 3 | 5 | 7 | 9 | 11 | 13 |
|---|---|---|---|---|---|---|---|
| $F_n/F$ | 1.2732 | 0.4244 | 0.2546 | 0.1819 | 0.1415 | 0.1157 | 0.0979 |
| (a) $X_n/\delta$ | 2.2096 | 0.0325 | 0.0067 | 0.0024 | 0.0011 | 0.0006 | 0.0004 |
| (b) $X_n/\delta$ | 1.3200 | 0.6224 | 1.7572 | 0.2406 | 0.0748 | 0.0349 | 0.0195 |

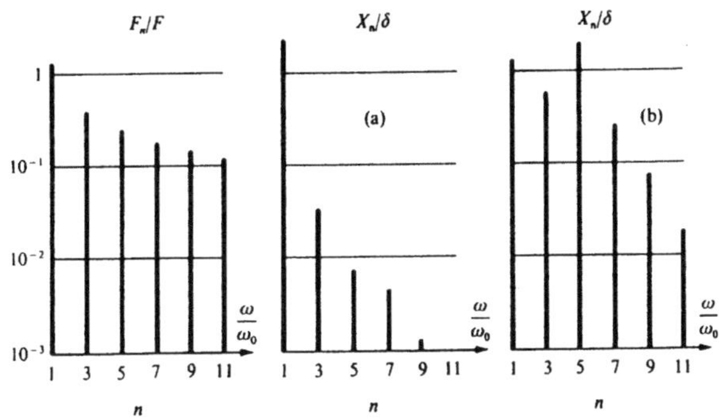


<u>Fig. 5.5</u> Summary table and presentation of the spectra f(t) and x(t) for example 5.3.2

<u>Comments</u>

- In practice it is impossible to produce a force identical to the one which we have chosen in this example. However, one can approximate to it closely by means of machines called hydropulsers, which are used for fatigue tests.

- The oscillator behaves like a very efficient low-pass filter in case (a). In effect, x(t) is a nearly perfect sinusoidal function since the amplitude of the most important harmonic - the third - amounts to only 0.0325/2.2096 = 1.5 % of the amplitude of the fundamental.

- In case (b), on the other hand, the 5th harmonic is close to the resonant amplitude $\omega_2 = \omega_0 \sqrt{1 - 2\,\eta^2} = 5.3\,\omega \sqrt{1 - 2 \times 0.05^2} = 5.29\,\omega$ . Its amplitude is thus bigger than that of the fundamental. Taking account of the 3rd and 7th harmonics as well, x(t) presents a shape significantly different from a pure sine wave.

### 5.3.3 Time response to a periodic excitation

Calculate the response of an elementary oscillator in the steady state acted on by the external force of figure 5.6, defined for one period $T = 2\,\pi/\omega$ by

$$f(t) = F(e^{-2t/T} - 1) \qquad -\frac{T}{2} \leqslant t \leqslant \frac{T}{2}$$

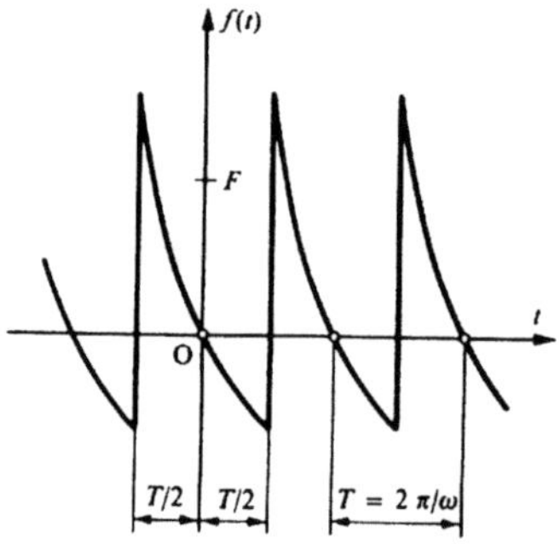


Fig. 5.6 Representation of the exciting force f(t)

The coefficients of the complex Fourier series of $f(t)$ are obtained by applying (5.13) to the period defined above

$$\underline{C}_n = \frac{1}{T}\int_{-T/2}^{T/2} f(t)\, e^{-jn\omega t}\, dt \tag{5.35}$$

then, by replacing $f(t)$ by its value

$$\underline{C}_n = F\,\frac{1}{2\,\pi}\int_{-\pi}^{\pi} (e^{-\frac{\omega t}{\pi}} - 1)\, e^{-jn\omega t}\, d(\omega t)$$

$$= F\,\frac{1}{2\,\pi}\int_{-\pi}^{\pi} (e^{-(1+jn\pi)\frac{\omega t}{\pi}} - e^{-jn\omega t})\, d(\omega t)$$

After integration, one obtains

$$\underline{C}_n = F\left(\frac{sh\ 1\ \cos n\pi}{1 + j\ n\ \pi} - \frac{\sin n\pi}{n\ \pi}\right) \tag{5.36}$$

In this way f(t) takes the simple form (5.12)

$$f(t) = \sum_{-\infty}^{\infty} \underline{C}_n\ e^{jn\omega t} \tag{5.37}$$

In this example, it is advantageous to express f(t) in the form (5.14)

$$f(t) = \mathrm{Re}\ \underline{f}(t) = \mathrm{Re}(D_0 + \sum_{n}^{\infty} \underline{D}_n\ e^{jn\omega t})$$

The coefficients $\underline{D}_n$ can be deduced directly from the $\underline{C}_n$ by (5.17). It gives here

$$D_0 = C_0 = F(sh\ 1 - 1)$$

$$\underline{D}_n = 2\ \underline{C}_n = 2\ F\left(\frac{sh\ 1\ \cos n\pi}{1 + j\ n\ \pi}\right) \tag{5.38}$$

These coefficients can be put into the exponential form

$$\underline{D}_n = 2\ F\left(\frac{sh\ 1\ \cos n\pi}{1 + j\ n\ \pi}\right) = 2\ F\ \frac{sh\ 1\ \cos n\pi}{\sqrt{1 + n^2\ \pi^2}}\ e^{-j\psi_n}$$

that is to say

$$\underline{D}_n = D_n\ e^{-j\psi_n} \tag{5.39}$$

with

$$D_n = 2\ F\ \frac{sh\ 1\ \cos n\pi}{\sqrt{1 + n^2\ \pi^2}} \qquad \mathrm{tg}\ \psi_n = n\pi$$

By using again the definitions (5.20) and (5.24), the response of the system is of the form (5.5)

$$x(t) = \bar{x} + \sum_{n}^{\infty} X_n \cos(n\omega t - \varphi_n - \psi_n) \tag{5.40}$$

where $\bar{x}$ represents the mean value of $x(t)$

$$\bar{x} = \frac{D_0}{k} = \frac{F}{k}\,(\text{sh } 1 - 1) = X_0(\text{sh } 1 - 1) \tag{5.41}$$

and

$$X_n = X_{sn}\,\mu_n = \frac{D_n}{k}\,\mu_n$$

$$X_n = 2\,X_0\,\frac{\text{sh } 1 \cos n\pi}{\sqrt{1 + n^2\,\pi^2}}\;\frac{1}{\sqrt{(1 - n^2\,\beta^2)^2 + 4\,\eta^2\,n^2\,\beta^2}} \tag{5.49}$$

In order to represent the external force and the response of the system as functions of time, the following values have been adopted

$$\beta = 0.26 \qquad \eta = 0.2 \qquad n = 20$$

and the amplitudes $F$ and $X_0$ have been chosen equal to unity.

Figure 5.7 allows one to compare $f(t)$, its decomposition into a Fourier series and the motion $x(t)$ of the system.

Comments

- The decomposition of $f(t)$ demonstrates the phenomenon, known as **Gibb's phenomenon,** which is circled in figure 5.7b. All Fourier series or integrals give birth to this phenomenon. To the left and right of a discontinuity, they cause oscillations whose amplitudes do not tend towards zero, when $n$ increases, but towards a value which is proportional to the jump at the discontinuity. The coefficient of proportionality has the value, from [36].

$$-\frac{1}{2} + \frac{1}{\pi}\int_0^{\pi} \frac{\sin u}{u}\,du \simeq 0,0895$$

The convergence of the series or integral is not affected by it due to the fact that, when $n$ becomes large, the space in which this phenomenon

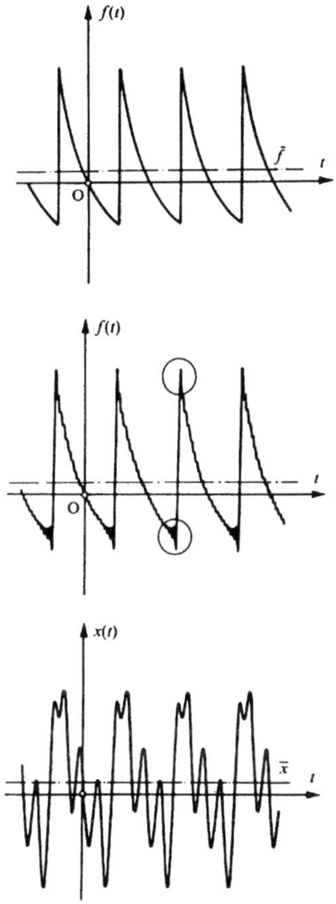


Fig. 5.7 The excitation f(t) and the reponse x(t) of the system plotted as a function of time

(a) f(t) according to its analytic definition

(b) Decomposition of f(t) into a Fourier series (n = 20)

(c) Response x(t) of the system for $\beta = 0.26 \quad \eta = 0.2$

is produced tends towards zero in the same region as the discontinuity, joining with it.

- For the value $\beta = 0.26$ $(1/4 < 0.26 < 1/3)$ chosen in this example, the 3rd and 4th harmonics predominate in the response x(t) , which is clearly visible in figure 5.7 (c).

- When $\beta$ is greater than 1, the response of the system is very close to a sine wave since the system behaves like a very efficient low-pass filter.

# CHAPTER 6 FORCED STATE

The forced state of an oscillator corresponds to the complete solution of the differential equation (2.1)

$$m \ddot{x} + c \dot{x} + k x = f(t)$$

It is therefore the sum of a specific solution of $x'(t)$ of the equation with the right-hand side not zero and of the general solution $x''(t)$ of the equation with the right-hand side equal to zero.

On the mathematical level, the problem of the forced state can be approached in various ways. Here are four possibilities :

- the direct search for specific solutions,
- the Laplace transformation,
- the Fourier transformation,
- numerical analysis.

Within the framework of this chapter, we shall limit ourselves to the Laplace and Fourier transforms (that of Fourier being a particular case of the Laplace transformation anyway).

## 6.1 Laplace transform

The Laplace transformation enables one to replace a differential calculus problem by one of algebra. We think it useful, in order to make it easier to read this chapter, to recall the essential properties and to give some elementary transforms in figure 6.1.

The **Laplace transform** of a function $f(t)$ of the real variable $t$, (this function being generated and defined for $t \geq 0$ ), is the integral.

$$F(s) = \int_0^\infty e^{-st} f(t)\, dt = L(f(t))$$

This transform is a function of the variable $s$, real or complex. The real number $\alpha$ which satisfies the condition $Re(s) > \alpha$ thereby ensuring the convergence of the integral, and thus the existence of $F(s)$, is called the **radius of convergence.**

(a)

| Fundamental properties of the Laplace transform | | |
|---|---|---|
| $f(t)$ | $F(s) = L(f(t))$ | Remarks |
| $C_1 f_1(t) + C_2 f_2(t)$ | $C_1 F_1(s) + C_2 F_2(s)$ | Linearity |
| $f'(t)$<br>$f''(t)$<br>$f^n(t)$ | $s\ F(s) - f(0)$<br>$s^2 F(s) - (s f(0) + f'(0))$<br>$s^n F(s) - (s^{n-1} f(0) + \cdots + s f^{n-2}(0) + f^{n-1}(0))$ | Transformation of derivatives |
| $h(t) = \int_0^t f(u)\, g(t-u)\, du = \int_0^t f(t-u)\, g(u)\, du$ | $H(s) = F(s)\, G(s)$ | Convolution integral |
| $g(t) = 0$ for $t < a$<br>$g(t) = f(t-a)$ for $t \geq a$ | $G(s) = e^{-as} F(s)$ | Displacement in the time domain (fig. 6.1 b) |
| $\int_{t_0}^t f(u)\, du$ | $\frac{1}{s} F(s) + \frac{1}{s} \int_{t_0}^{0} f(u)\, du$ | Transformation |
| $\int_0^t f(u)\, du$ | $\frac{1}{s} F(s)$ | of integrals |
| $e^{-at} f(t)$ | $F(s+a)$ | Displacement in the transform domain |

Fig. 6.1 Laplace Transform
(a) Table of fundamental properties
(b) Delay of the function $g(t)$ with respect to the function $f(t)$
(c) Table of elementary transforms

(b)

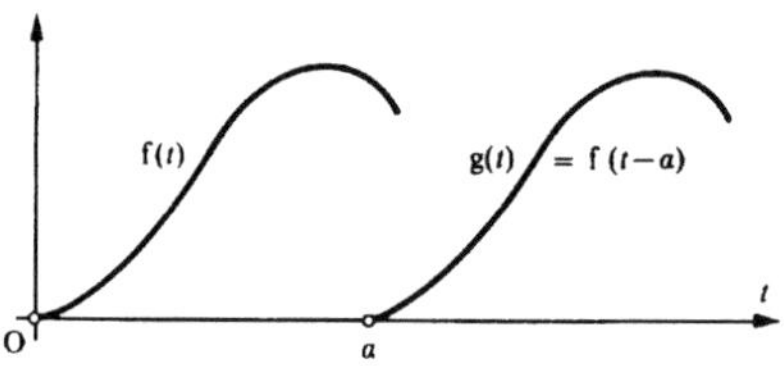


(c)

| Table of Elementary Laplace transforms | | | |
|---|---|---|---|
| f(t) | F(s) = L(f(t)) | f(t) | F(s) = L(f(t)) |
| $1$ | $\frac{1}{s}$ | $t \cos \omega t$ | $\frac{s^2-\omega^2}{(s^2+\omega^2)^2}$ |
| $t$ | $\frac{1}{s^2}$ | $t \sin \omega t$ | $\frac{2 \omega s}{(s^2+\omega^2)^2}$ |
| $t^n$ | $\frac{n!}{s^{n+1}}$ | $e^{-at} \cos \omega t$ | $\frac{s + a}{(s+a)^2+\omega^2}$ |
| $e^{at}$ | $\frac{1}{s-a}$ | $e^{-at} \sin \omega t$ | $\frac{\omega}{(s+a)^2+\omega^2}$ |
| $t\ e^{at}$ | $\frac{1}{(s-a)^2}$ | $ch\ \omega t$ | $\frac{s}{s^2-\omega^2}$ |
| $t^n\ e^{at}$ | $\frac{n!}{(s-a)^{n+1}}$ | $sh\ \omega t$ | $\frac{\omega}{s^2-\omega^2}$ |
| $(1+at)\ e^{at}$ | $\frac{s}{(s-a)^2}$ | $1 - \cos \omega t$ | $\frac{\omega^2}{s(s^2+\omega^2)}$ |
| $\frac{1}{r_1-r_2}(e^{r_1 t}-e^{r_2 t})$ | $\frac{1}{(s-r_1)(s-r_2)}$ | $ch\ \omega t - 1$ | $\frac{\omega^2}{s(s^2-\omega^2)}$ |
| $\cos \omega t$ | $\frac{s}{s^2+\omega^2}$ | $\sqrt{t}$ | $\frac{\sqrt{\pi}}{2 s \sqrt{s}}$ |
| $\sin \omega t$ | $\frac{\omega}{s^2+\omega^2}$ | $\frac{1}{\sqrt{t}}$ | $\frac{\sqrt{\pi}}{\sqrt{s}}$ |

It is necessary to point out that it is also used in a modified form, known as the **Carson-Laplace transform**, which is defined as follows

$$F'(s) = s \int_0^\infty e^{-st} f(t)\, dt = L'(f(t))$$

This has the advantage of transforming a constant into a constant. In order to use the tables, it is sufficient to remember the obvious relations

$$F'(s) = s\,F(s) \quad \Rightarrow \quad F(s) = \frac{1}{s}\,F'(s)$$

6.2 General solution of the forced state

Let us go back to equation (2.1)

$$m\,\ddot{x} + c\,\dot{x} + k\,x = f(t)$$

and let us take the Laplace transform of the two sides.

With the notation

$$X(s) = L(x(t)) \qquad x(0) = X_0 \qquad \dot{x}(0) = V_0$$

$$F(s) = L(f(t))$$

it gives

$$m(s^2\,X(s) - s\,X_0 - V_0) + c(s\,X(s) - X_0) + k\,X(s) = F(s)$$

$$X(s)\,(m\,s^2 + c\,s + k) = F(s) + X_0\,(m\,s + c) + V_0\,m \tag{6.1}$$

The following quantity is referred to by the name **operational impedance**

$$Z(s) = m\,s^2 + c\,s + k \tag{6.2}$$

Its inverse is called the **operational admittance** or **transfer function**

$$Y(s) = \frac{1}{m\,s^2 + c\,s + k} \tag{6.3}$$

Before continuing, it is convenient to write $Z(s)$ as follows

$$Z(s) = m(s^2 + \frac{c}{m} s + \frac{k}{m})$$

One introduces the quantities defined by the relations (2.2) to (2.4) as follows

$$\omega_0^2 = \frac{k}{m} \qquad \lambda = \frac{c}{2\,m} \qquad \eta = \frac{\lambda}{\omega_0}$$

Thus

$$Z(s) = m(s^2 + 2\,\lambda\,s + \omega_0^2) \tag{6.4}$$

$$Y(s) = \frac{1}{m}\,\frac{1}{s^2 + 2\,\lambda\,s + \omega_0^2} \tag{6.5}$$

The transform of $X(s)$ of the solution sought $x(t)$ has the form, may be (6.1) and (6.5)

$$X(s) = Y(s)\,F(s) + \frac{X_0(s + 2\,\lambda) + V_0}{s^2 + 2\,\lambda\,s + \omega_0^2} \tag{6.6}$$

If the inverse functions of the two terms on the right-hand side are $x_a(t)$ and $x_b(t)$ respectively, it gives

$$x(t) = x_a(t) + x_b(t)$$

The function $x_b(t)$ is known. After having determined the **temporal admittance** $\mathbf{y(t)}$, being the inverse of $Y(s)$, $x_a(t)$ may be calculated by means of the **convolution integral**.

$$x_1(t) = \int_0^t y(t-u)\,f(u)\,du = \int_0^t y(u)\,f(t-u)\,du \tag{6.7}$$

As before, it is necessary to distinguish three cases as a function of the value of the damping factor $\eta$ .

Super-critical damping $\eta > 1$

$$\eta > 1 \quad \Rightarrow \quad \lambda^2 > \omega_0^2 \quad \Rightarrow \quad \omega_1^2 = \lambda^2 - \omega_0^2 \qquad \text{(relation 3.35)}$$

The admittance can be put in the form

$$Y(s) = \frac{1}{m} \frac{1}{s^2 + 2\lambda s + \omega_0^2} = \frac{1}{m} \frac{1}{(s - r_1)(s - r_2)}$$

with

$$\begin{cases} r_1 = -\lambda + \omega_1 \\ r_2 = -\lambda - \omega_1 \end{cases}$$

where, by using the table of transforms

$$y(t) = \frac{e^{-\lambda t}}{2 m \omega_1} (e^{\omega_1 t} - e^{-\omega_1 t}) = \frac{e^{-\lambda t}}{m \omega_1} \operatorname{sh}\omega_1 t \qquad (6.8)$$

For $x_b(t)$, the most convenient is to use relation (3.41). The forced state is thus

$$x(t) = \frac{1}{m \omega_1} \int_0^t e^{-\lambda u} \operatorname{sh}\omega_1 u \cdot f(t - u)\, du + e^{-\lambda t} \left(X_0 \operatorname{ch}\omega_1 t + \frac{\lambda X_0 + V_0}{\omega_1} \operatorname{sh}\omega_1 t\right) \qquad (6.9)$$

Critical damping $\eta = 1$

$$\eta = 1 \quad \Rightarrow \quad \lambda = \omega_0$$

The admittance becomes

$$Y(s) = \frac{1}{m} \frac{1}{s^2 + 2\omega_0 s + \omega_0^2} = \frac{1}{m} \frac{1}{(s + \omega_0)^2}$$

As a result, using the table

$$y(t) = \frac{t}{m} e^{-\omega_0 t} \qquad (6.10)$$

The function $x_b(t)$ being given by (3.47) , one has

$$x(t) = \frac{1}{m}\int_0^t u\, e^{-\omega_0 u}\, f(t-u)\, du + (X_0 + (\omega_0 X_0 + V_0)\, t)\, e^{-\omega_0 t} \qquad (6.11)$$

Sub-critical damping $\eta < 1$

$$\eta < 1 \quad \Rightarrow \quad \lambda^2 < \omega_0^2 \quad \Rightarrow \quad \omega_1^2 = \omega_0^2 - \lambda^2$$

The most convenient way to use the table is to proceed as follows

$$Y(s) = \frac{1}{m}\,\frac{1}{s^2 + 2\lambda s + \omega_0^2} = \frac{1}{m}\,\frac{1}{(s+\lambda)^2 + \omega_0^2 - \lambda^2} = \frac{1}{m\,\omega_1}\,\frac{\omega_1}{(s+\lambda)^2 + \omega_1^2}$$

$$y(t) = \frac{e^{-\lambda t}}{m\,\omega_1}\sin\omega_1 t \qquad (6.12)$$

This result, just like the relations (3.57) and (3.58), determines the forced state

$$x(t) = \frac{1}{m\,\omega_1}\int_0^t e^{-\lambda u}\sin\omega_1 u \cdot f(t-u)\, du + X\, e^{-\lambda t}\cos(\omega_1 t - \varphi) \qquad (6.13)$$

with

$$X = \sqrt{X_0^2 + \left(\frac{\lambda X_0 + V_0}{\omega_1}\right)^2} \qquad \operatorname{tg}\varphi = \frac{\lambda X_0 + V_0}{\omega_1 X_0}$$

The integrals appearing in the above relations (specific cases of the integral **of** Duhamel) are the basis of certain methods of numerical analysis which we will not consider here. When the external force $f(t)$ is defined by an analytic function, it is nearly always preferable to replace the calculation of these integrals by the direct inversion of the product $Y(s)\,F(s)$, using the table of transforms.

As examples, we are now going to calculate the forced states stimulated by two typical exciting forces

· an impulse force or displacement,
· a unit step force or displacement.

These states are called the **impulse response** and the **indicial response** respectively. We will assume that the initial conditions are zero. If that is not the case, it is sufficient to put in the correctly chosen function $x_b(t)$ (free state).

## 6.3 Response to an impulse and to a unit step force

### 6.3.1 Impulse response

Let us suppose that a very sudden (impulse-like) external force F is applied to the oscillator for a very short interval of time $\varepsilon$ (figure 6.2). By analogy with the Dirac delta function, one such stimulation is called a **Dirac impulse** if the product $F\,\varepsilon$ is equal to unity when $\varepsilon \to 0$ and $F \to \infty$ .

$$F\,\varepsilon = 1 \qquad [F\,\varepsilon] = 1\ \mathrm{N\ s} = 1\ \text{Newton x second}$$

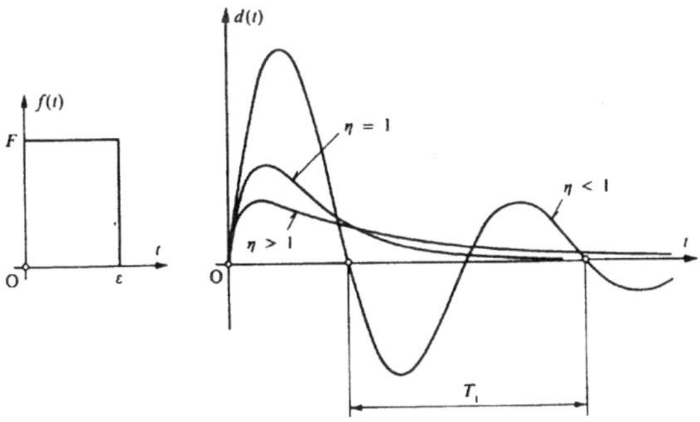


Fig. 6.2 Dirac impulse and impulse response

One shows that the Laplace transform of the Dirac impulse is equal to 1. Consequently, if one designates the response of the oscillator by $d(t)$, known as the **impulse response** , one will have from (6.6)

$$D(s) = Y(s) \cdot 1 \quad \Rightarrow \quad d(t) = y(t) \tag{6.14}$$

This response is thus particularly simple and interesting; it is equal to the temporal admittance defined and calculated in section 6.2. Moreover, before continuing, it is necessary to observe that the physical dimensions of these quantities are different. In effect, if the displacement x(t) is a length, one establishes by dimensional analysis the following (m = metre, kg = kilogram, s = second, N = newton, J = joule) :

- impulse response — $[d(t)] = m$
- temporal admittance — $[y(t)] = s/kg$
- operational admittance — $[Y(s)] = N/m = s^2/kg$
- Laplace variable — $[s] = [\omega_0] = rad/s$ (whatever the value of x(t))

Having noted this, one can directly use the preceding results.

<u>Super-critical damping</u> $\eta > 1$

$$d(t) = \frac{e^{-\lambda t}}{2\, m\, \omega_1}\left(e^{\omega_1 t} - e^{-\omega_1 t}\right) = \frac{e^{-\lambda t}}{m\, \omega_1}\, \mathrm{sh}\,\omega_1 t \tag{6.15}$$

<u>Critical damping</u> $\eta = 1$

$$d(t) = \frac{t}{m}\, e^{-\omega_0 t} \tag{6.16}$$

<u>Sub-critical damping</u> $\eta < 1$

$$d(t) = \frac{e^{-\lambda t}}{m\, \omega_1} \sin \omega_1 t \tag{6.17}$$

Figure 6.2 shows d(t) for three values of the damping factor. It is easy to prove that the three curves have a common tangent at the origin

$$\dot{d}(0) = \frac{1}{m} \qquad [\dot{d}(0)] = \frac{N \cdot s}{kg} = \frac{kg \cdot m \cdot s}{s^2 \cdot kg} = m/s \tag{6.18}$$

So the velocity of the mass passes abruptly from zero to 1/m , whatever the damping. This discontinuity in the velocity requires an infinite acceleration, which in turn requires an infinite force. This is actually the case for a Dirac impulse, but in an interval $0 \leqslant t \leqslant \varepsilon$ of null duration (one cannot check such an assertion by calculating d(t) from the preceding relations which are only valid for $t > \varepsilon$ ).

The energy initially supplied to the system is all in the kinetic form. It has the value

$$T(0) = \frac{1}{2}\, m\, \dot{d}^2(0) = \frac{1}{2\, m} \qquad [T(0)] = kg \cdot \left(\frac{m}{s}\right)^2 = N \cdot m = \text{joule} \tag{6.19}$$

It is therefore equal to 0.5 joule when the mass of the oscillator is 1 kg.

To summarize, the response d(t) of the oscillator to a Dirac impulse corresponds to the temporal admittance y(t) as well as to the free state with the initial conditions $X_0 = 0$ and $V_0 = 1/m$ .

### 6.3.2 Indicial response

An external force defined as follows is called a **unit step force** (figure 6.3)

$$f(t) = 0 \qquad \text{for} \quad t < 0$$
$$f(t) = 1 \qquad \text{for} \quad t \geqslant 0$$

Its Laplace transform is therefore $F(s) = 1/s$ . By designating the corresponding response of the oscillator, the said indicial response, by $e(t)$ , relation (6.6) gives

$$E(s) = \frac{1}{s} Y(s) \tag{6.20}$$

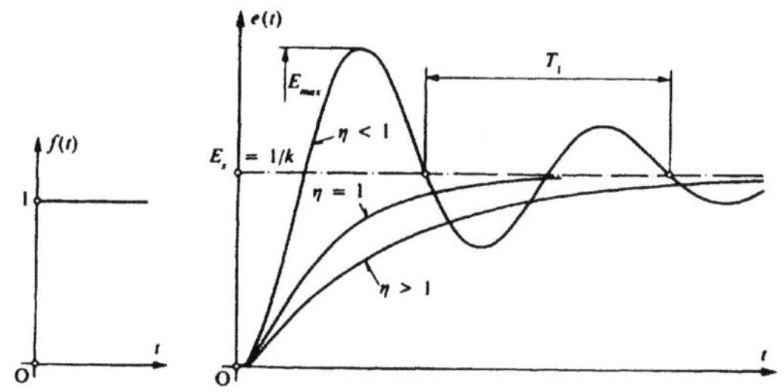


Fig. 6.3 Unit step force and indicial response

Super-critical damping $\qquad \eta > 1 \qquad \Rightarrow \qquad \omega_1^2 = \lambda^2 - \omega_0^2$

$$E(s) = \frac{1}{m} \frac{1}{s(s - r_1)(s - r_2)} \tag{6.21}$$

with

$$\begin{cases} r_1 = -\lambda + \omega_1 \\ r_2 = -\lambda - \omega_1 \end{cases}$$

One reduces the fraction to simple terms

$$E(s) = \frac{1}{m} \left(\frac{\alpha}{s} + \frac{\beta}{s - r_1} + \frac{\gamma}{s - r_2}\right) \quad \Rightarrow \quad e(t) = \frac{1}{m} (\alpha + \beta\, e^{r_1 t} + \gamma\, e^{r_2 t})$$

The identification gives the three following equations

$$\begin{cases} 0 = \alpha + \beta + \gamma \\ 0 = \alpha(r_1 + r_2) + \beta\, r_2 + \gamma\, r_1 \\ 1 = \alpha\, r_1\, r_2 \end{cases}$$

which have as solutions

$$\begin{cases} \alpha = \dfrac{1}{r_1\ r_2} = \dfrac{1}{\omega_0^2} = \dfrac{m}{k} \\ \beta = \dfrac{1}{r_1(r_1 - r_2)} = \dfrac{-1}{2\ \omega_1(\lambda - \omega_1)} \\ \gamma = \dfrac{-1}{r_2(r_1 - r_2)} = \dfrac{1}{2\ \omega_1(\lambda + \omega_1)} \end{cases}$$

The response of the oscillator is thus

$$e(t) = \frac{1}{k} - \frac{e^{-\lambda t}}{2\ m\ \omega_1}\left(\frac{e^{\omega_1 t}}{\lambda - \omega_1} - \frac{e^{-\omega_1 t}}{\lambda + \omega_1}\right)$$

since

$$(\lambda - \omega_1)(\lambda + \omega_1) = \omega_0^2 = \frac{k}{m}$$

it can be put in the form

$$e(t) = \frac{1}{k}\left(1 - e^{-\lambda t}\left(ch\omega_1 t + \frac{\lambda}{\omega_1} sh\omega_1 t\right)\right) \qquad (6.22)$$

This is a non-periodic function which tends towards an horizontal asymptote corresponding to the static displacement (figure 6.3).

$$E_s = \frac{1}{k} \qquad [E_s] = \frac{N}{N/m} = m \qquad (6.23)$$

Critical damping $\eta = 1 \quad \Rightarrow \quad \lambda = \omega_0$

$$E(s) = \frac{1}{m}\frac{1}{s(s+\omega_0)^2} \qquad (6.24)$$

One has successively

$$E(s) = \frac{1}{m}\left(\frac{\alpha}{s} + \frac{\beta}{(s+\omega_0)^2} + \frac{\gamma s}{(s+\omega_0)^2}\right)$$

$$e(t) = \frac{1}{m}\left(\alpha + \beta\, t\, e^{-\omega_0 t} + \gamma(1 - \omega_0 t)\, e^{-\omega_0 t}\right)$$

$$\begin{cases} 0 = \alpha + \gamma \\ 0 = 2\,\alpha\,\omega_0 + \beta \\ 1 = \alpha\,\omega_0^2 \end{cases} \Rightarrow \begin{cases} \alpha = \dfrac{1}{\omega_0^2} \\ \beta = \dfrac{-2}{\omega_0} \\ \gamma = \dfrac{-1}{\omega_0^2} \end{cases}$$

$$e(t) = \frac{1}{k}\left(1 - (1 + \omega_0 t)\, e^{-\omega_0 t}\right) \qquad (6.25)$$

This function has the same shape as the preceding.

3) <u>Sub-critical damping</u> $\eta < 1 \Rightarrow \omega_1^2 = \omega_0^2 - \lambda^2$

$$E(s) = \frac{1}{m}\,\frac{1}{s((s+\lambda)^2 + \omega_1^2)} \qquad (6.26)$$

One can proceed as follows

$$E(s) = \frac{1}{m}\left(\frac{\alpha}{s} + \frac{\beta + \gamma s}{(s+\lambda)^2 + \omega_1^2}\right) = \frac{1}{m}\left(\frac{\alpha}{s} + \gamma\,\frac{s+\lambda}{(s+\lambda)^2 + \omega_1^2} + \right.$$

$$\left. + \frac{\beta - \gamma\lambda}{\omega_1}\,\frac{\omega_1}{(s+\lambda)^2 + \omega_1^2}\right)$$

$$e(t) = \frac{1}{m}\left(\alpha + \gamma\, e^{-\lambda t} \cos \omega_1 t + \frac{\beta - \gamma\lambda}{\omega_1}\, e^{-\lambda t} \sin \omega_1 t\right)$$

$$\begin{cases} 0 = \alpha + \gamma \\ 0 = 2\,\alpha\,\lambda + \beta \\ 1 = \alpha(\lambda^2 + \omega^2) \end{cases} \Rightarrow \begin{cases} \alpha = \dfrac{1}{\lambda^2 + \omega_1^2} = \dfrac{1}{\omega_0^2} = \dfrac{m}{k} \\ \beta = -\,2\,\alpha\,\lambda = \dfrac{-2\,m\,\lambda}{k} \\ \gamma = -\,\dfrac{m}{k} \end{cases}$$

$$e(t) = \frac{1}{k} (1 - e^{-\lambda t} (\cos \omega_1 t + \frac{\lambda}{\omega_1} \sin \omega_1 t)) \tag{6.27}$$

The displacement $e(t)$ undergoes decreasing oscillations about the equilibrium position $E_s$.

When this position is reached - after an infinite time in principle - and whatever the damping factor, the external force has supplied work to the system.

$$H_\infty = 1 \cdot E_s = \frac{1}{k} \qquad [H] = N \cdot m = \text{joule} \tag{6.28}$$

while the accumulated potential energy has the value

$$V_\infty = \frac{1}{2} k E_s^2 = \frac{1}{2k} \tag{6.29}$$

This energy, which is equal to 0.5 joule when the stiffness is 1N/m, thus represents half of the work of the external force. The other half has therefore been dissipated by the resistance $c$.

To summarize, in the case of the response $e(t)$ to a unit step force, the oscillator tends towards a static equilibrium position $E_s = 1/k$ ; half of the work supplied is stored whilst the other half is lost.

If the external force is a step $F_0$ instead of a unit step, it is sufficient to multiply the relations (6.22), (6.25) and (6.27) by $F_0$ because of the linearity of the system. As for the final potential energy, it becomes

$$V_\infty(F_0) = (\frac{F_0}{1})^2 V_\infty = \frac{F_0^2}{2k} \tag{6.30}$$

### 6.3.3 Relation between the impulse and indicial responses

The response $d(t)$ to a Dirac impulse is the derivative with respect to time of the response $e(t)$ to a step force. In order to demonstrate this, let us write the relation (6.6) for these two responses

$$D(s) = Y(s) \cdot 1$$

$$E(s) = Y(s) \cdot \frac{1}{s}$$

One has then

$$D(s) = s\,E(s) \tag{6.31}$$

Let us take the Laplace transform of $\dot{e}(t) = \frac{de}{dt}$

$$L(\dot{e}(t)) = s\,E(s) \qquad (e(0) = 0) \tag{6.32}$$

It gives, by comparing (6.31) and (6.32)

$$D(s) = L(\dot{e}(t)) \tag{6.33}$$

and consequently

$$d(t) = \dot{e}(t) \tag{6.34}$$

## 6.4 Responses to an impulse and to a unit step elastic displacement

### 6.4.1 Introduction

One often encounters a slightly different conception of the indicial and impulse responses from that presented in the previous section. These are the responses of the oscillator to an impulse and to a unit step of the **elastic displacement** $x_e(t)$ respectively. Let us recall that $x_e(t)$ is the displacement which the external force $f(t)$ would cause on a system involving only the stiffness $k$ (relation (2.6))

$$x_e(t) = \frac{1}{k}\,f(t)$$

The initial conditions being assumed null, let us go back to relation (6.6)

$$X(s) = k\,Y(s) \cdot \frac{1}{k}\,F(s) = k\,Y(s) \cdot X_e(s)$$

with

$$X_e(s) = L(x_e(t))$$

By analogy with the frequency response $\underline{H}(\omega)$ defined by the relation (4.24), one gives the name **operational response** to the product $k\,Y(s)$, which gives using (6.5)

$$H(s) = \frac{k}{m}\,\frac{1}{s^2 + 2\,\lambda\,s + \omega_0^2} = \omega_0^2\,\frac{1}{s^2 + 2\,\lambda\,s + \omega_0^2}$$

$$H(s) = \frac{1}{\dfrac{s^2}{\omega_0^2} + \dfrac{2\,\lambda\,s}{\omega_0^2} + 1} = \frac{1}{\dfrac{s^2}{\omega_0^2} + \dfrac{2\,\eta\,s}{\omega_0} + 1} \qquad (6.35)$$

Before continuing, one can make the following remarks :

- the quantity $H(s)$ appears by taking the Laplace transform of the relation (2.7);
- if $s = j\omega$, $H(s)$ becomes $\underline{H}(\omega)$;
- $H(s)$ has no physical dimension.

### 6.4.2 Impulse response

If the displacement $x_e(t)$ is a Dirac impulse, $X_e(s) = 1$ and $X(s) = H(s)$. The corresponding function of time $h(t)$ is the impulse response to an impulse elastic displacement. One determines it directly from relations (6.15) to (6.17), taking into account (3.35).

Super-critical damping $\eta > 1$

$$h(t) = \frac{\omega_0}{2\sqrt{\eta^2 - 1}} e^{-\lambda t} (e^{\omega_1 t} - e^{-\omega_1 t}) = \frac{\omega_0}{\sqrt{\eta^2 - 1}} e^{-\lambda t} \operatorname{sh}\omega_1 t \qquad (6.36)$$

Critical damping $\eta = 1$

$$h(t) = \omega_0^2 \, t \, e^{-\omega_0 t} \qquad (6.37)$$

Sub-critical damping $\eta < 1$

$$h(t) = \frac{\omega_0}{\sqrt{1 - \eta^2}} e^{-\lambda t} \sin \omega_1 t \qquad (6.38)$$

The curves $h(t)$ have exactly the same shape as the curves $d(t)$ of figure 6.2.

### 6.4.3 Indicial response

When the displacement $x_e(t)$ is a unit step, $X(s) = \frac{1}{s}$ and $X(s) = H(s)/s$ . The function of time $g(t)$ is then the indicial response to a step elastic displacement. It can be determined from relations (6.22), (6.25) and (6.27) which it suffices to multiply by the stiffness $k$ . The corresponding curves have the same shape as those of figure 6.3.

Super-critical damping $\eta > 1$

$$g(t) = 1 - e^{-\lambda t}(\operatorname{ch}\omega_1 t + \frac{\lambda}{\omega_1} \operatorname{sh}\omega_1 t) \qquad (6.39)$$

Critical damping $\eta = 1$

$$g(t) = 1 - (1 + \omega_0 t)\, e^{-\omega_0 t} \tag{6.40}$$

Sub-critical damping $\eta < 1$

$$g(t) = 1 - e^{-\lambda t}(\cos \omega_1 t + \frac{\lambda}{\omega_1} \sin \omega_1 t) \tag{6.41}$$

## 6.5 Fourier transformation

This transformation, the most commonly used in vibratory analysis, is often presented as a particular case of the Laplace transform, where the Laplace variable $s$ is purely imaginary and written $j\omega$.

We think it preferable to establish it here as an extension of the Fourier series to the decomposition of non-periodic functions.

We have seen in chapter 5 that a periodic function, of period T, can be represented by a Fourier series, that is to say by an infinite series of harmonic functions of angular frequency $n\omega$ ($n = 0, \pm 1, \pm 2, \ldots$) where $\omega = 2\pi/T$ is the angular frequency of the fundamental. If one stretches the period $T$ towards infinity, in such a way that the first interval of time increases indefinitely, the function becomes non-periodic. In this process, the discrete frequencies draw ever closer to one another until they constitute a continuous spectrum. At that moment, the Fourier series becomes the **Fourier integral** .

Let us use again the periodic function of figure 5.1 which was expressed by the complex form of the Fourier series (5.12)

$$f(t) = \sum_{n=-\infty}^{\infty} \underline{c}_n\, e^{jn\omega t} \qquad \omega = 2\pi/T$$

and therefore the coefficients $\underline{C}_n$ were obtained from (5.13)

$$\underline{C}_n = \frac{1}{T} \int_0^T f(t)\, e^{-jn\omega t}\, dt \qquad n = 0, \pm 1, \pm 2, \ldots$$

By adopting the following notation

$$n\,\omega = \omega_n$$

$$(n+1)\,\omega - n\,\omega = \omega = 2\pi/T = \Delta\omega_n$$

we can put the preceding relations in the form

$$f(t) = \sum_{-\infty}^{\infty} \frac{1}{T} (T\,\underline{C}_n)\, e^{j\omega_n t} = \frac{1}{2\pi} \sum_{-\infty}^{\infty} (T\,\underline{C}_n)\, e^{j\omega_n t}\, \Delta\omega \qquad (6.42)$$

$$T\,\underline{C}_n = \int_0^T f(t)\, e^{-j\omega_n t}\, dt = \int_{-T/2}^{T/2} f(t)\, e^{-j\omega_n t}\, dt \qquad (6.43)$$

When the period increases indefinitely, $T \to \infty$, the index $n$ can be eliminated and the discrete variable $\omega_n$ becomes the continuous variable $\omega$. After going to the limit, the summation is replaced by an integration and one obtains

$$f(t) = \lim_{\substack{T\to\infty \\ \Delta\omega_n\to 0}} \frac{1}{2\pi} \sum_{-\infty}^{\infty} (T\,\underline{C}_n)\, e^{j\omega_n t}\, \Delta\omega_n = \frac{1}{2\pi} \int_{-\infty}^{\infty} \underline{F}(\omega)\, e^{j\omega t}\, d\omega \qquad (6.44)$$

$$\underline{F}(\omega) = \lim_{\substack{T\to\infty \\ \Delta\omega_n\to 0}} (T\,\underline{C}_n) = \int_{-\infty}^{\infty} f(t)\, e^{-j\omega t}\, dt \qquad (6.45)$$

The relation (6.44) expresses the fact that any function $f(t)$ can be described by an integral representing the contributions of the harmonic components having a continuous frequency spectrum from $-\infty$ to $+\infty$.

Equation (6.45) defines the Fourier transform $\underline{F}(\omega)$ of the function of time $f(t)$ . One can say that the quantity $\underline{F}(\omega)d\omega$ represents the contribution to $f(t)$ of the harmonics included in the frequency region from $\omega$ to $\omega + d\omega$.

The integrals

$$\underline{F}(\omega) = \int_{-\infty}^{\infty} f(t)\, e^{-j\omega t}\, dt \qquad (6.46)$$

and

$$f(t) = \frac{1}{2\pi} \int_{-\infty}^{\infty} F(\omega)\, e^{j\omega t}\, d\omega \qquad (6.47)$$

constitute a **pair of transforms** in which $f(t)$ is called the **inverse Fourier transform** of $\underline{F}(\omega)$.

By analogy with the decomposition using the Fourier series, the relations (6.46) and (6.47) give the frequency composition of the non-periodic function $f(t)$.

The representation of $f(t)$ by the integral (6.47) is only possible if the integral (6.46) exists. For that, $f(t)$ must satisfy the condition of Dirichlet[1] in the time range $-\infty < t < \infty$ and the integral

$$I = \int_{-\infty}^{\infty} |f(t)|\, dt \qquad (6.48)$$

must be convergent. In the case where this diverges, the Fourier transform

---

(1) The function $f(t)$ would satisfy the condition of Dirichlet in the interval $(a,b)$ if :
- $f(t)$ has only a finite number of maxima and minima in $(a,b)$ and
- $f(t)$ has only a finite number of discontinuities in the interval $(a,b)$ and not any infinite discontinuities

$\underline{F}(\omega)$ does not exist and the function can generally be handled by the Laplace transform (with the condition that this transform is itself convergent).

It is easy to see that if the excitation $f(t)$ is put in the form (5.12), the complex response $\underline{x}(t)$ of the oscillator becomes, from (5.17) and (5.19)

$$\underline{x}(t) = \sum_{-\infty}^{\infty} \underline{Y}_n \underline{C}_n e^{jn\omega t} = \frac{1}{k} \sum_{-\infty}^{\infty} \underline{H}_n \underline{C}_n e^{jn\omega t} \tag{6.49}$$

In this relation, $\underline{Y}_n$ and $\underline{H}_n$ are respectively the complex admittance and the complex frequency response relative to the angular frequency $n\omega$.

By proceeding in the same way as for $f(t)$, we can write the complex response of the system to any non-periodic excitation in the form of a pair of Fourier transforms

$$\underline{X}(\omega) = \int_{-\infty}^{\infty} \underline{x}(t)\, e^{-j\omega t}\, dt \tag{6.50}$$

$$\underline{x}(t) = \frac{1}{2\pi} \int_{-\infty}^{\infty} \underline{X}(\omega)\, e^{j\omega t}\, d\omega \tag{6.51}$$

The transforms of the complex response and of the excitation are linked by the condition

$$\underline{X}(\omega) = \underline{Y}(\omega)\, \underline{F}(\omega) = \frac{1}{k} \underline{H}(\omega)\, \underline{F}(\omega) \tag{6.52}$$

## 6.6 Examples of forced states

### 6.6.1 Time response to a force $F \cos \omega t$

Calculate, by means of the Laplace transform, the forced state of an elementary oscillator to which one applies the external force $f(t) = F \cos \omega t$, at time $t = 0$. One assumes that the initial conditions are zero : $X_0 = 0$, $V_0 = 0$.

One uses the relations (6.6) and (6.5)

$$X(s) = Y(s)\, F(s) \qquad\qquad Y(s) = \frac{1}{m} \frac{1}{s^2 + 2\lambda s + \omega_0^2}$$

In this particular case

$$F(s) = L(F \cos \omega t) = F \frac{s}{s^2 + \omega^2}$$

and consequently

$$X(s) = \frac{F}{m} \frac{s}{(s^2 + \omega^2)(s^2 + 2\lambda s + \omega_0^2)} \tag{6.53}$$

After decomposition into simple elements and inversion, it gives, after all calculations are done

$$x(t) = X(\cos(\omega t - \varphi) - \xi e^{-\lambda t} \cos(\omega_1 t - \varphi_1)) \tag{6.54}$$

with

$$\xi = \mu \sqrt{(1 - \beta^2)^2 + \frac{\eta^2}{1 - \eta^2}(1 + \beta^2)^2} \qquad\qquad \operatorname{tg} \varphi_1 = \frac{\lambda}{\omega_1} \frac{1 + \beta^2}{1 - \beta^2} \tag{6.55}$$

As for the other symbols, they have the usual meaning

$$X = \mu \frac{F}{k} \qquad \omega_0^2 = \frac{k}{m} \qquad \eta = \frac{\lambda}{\omega_0} \qquad \beta = \frac{\omega}{\omega_0}$$

$$\omega_1^2 = \omega_0^2(1 - \eta^2) \qquad \mu = \frac{1}{\sqrt{(1 - \beta^2)^2 + 4\eta^2\beta^2}} \qquad \operatorname{tg} \varphi = \frac{2\eta\beta}{1 - \beta^2}$$

The function $x(t)$ can have significantly different shapes as a function of the values of the principal parameters of the problem, which are the relative angular frequency $\beta$ and the damping factor $\eta$ . Let us take the following example

$$f = 5 \text{ Hz} \qquad f_0 = 17 \text{ Hz} \qquad \eta = 0{,}05 \qquad \mu \frac{F}{k} = X = 1$$

From these values, one finds

$$\omega = 2\pi f = 31{,}4 \text{ s}^{-1} \qquad \omega_0 = 2\pi f_0 = 106{,}8 \text{ s}^{-1} \qquad \omega_1 = 106{,}7 \text{ s}^{-1}$$

$$\beta = \frac{5}{17} = 0{,}294 \qquad \lambda = 5{,}34 \text{ s}^{-1}$$

$$\mu = 1{,}094 \qquad \xi = 1{,}0013 \qquad \varphi = 0{,}032 \qquad \varphi_1 = 0{,}060$$

The forced state is now determined (figure 6.4)

$$x(t) = \cos(31{,}4\, t - 0{,}032) - 1{,}0013\, e^{-5{,}34\, t} \cos(106{,}7\, t - 0{,}060)$$

$$x(t) = x'(t) + x''(t)$$

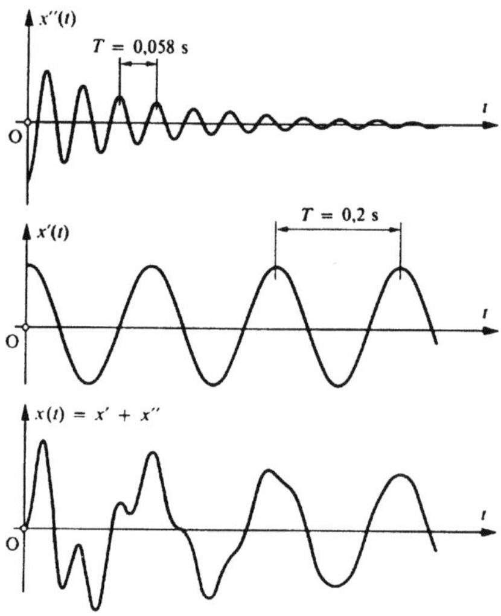


Fig. 6.4 Forced state due to a force F cos ωt

It includes the steady state $x'(t)$, of angular frequency $\omega$, as well as a transitory term $x''(t)$ which oscillates at the natural angular frequency $\omega_1$ and which would practically disappear after about 15 periods.

Let us recall that the same problem can be solved without using the Laplace transform. In fact, the solution of the equation of the oscillator

$$m\ddot{x} + c\dot{x} + kx = F\cos\omega t$$

is the sum of the general solution $x''(t)$ of the equation without the right-hand side and of a specific solution $x'(t)$ of the complete equation

$$x(t) = x'(t) + x''(t) \tag{6.56}$$

The relations (4.4) to (4.6) determine

$$x'(t) = X\cos(\omega t - \varphi)$$

$$\text{with } X = \frac{F}{\sqrt{(k - \omega^2 m)^2 + \omega^2 c^2}} \qquad \operatorname{tg}\varphi = \frac{\omega c}{k - \omega^2 m} \tag{6.57}$$

The relation (3.57) gives $x''(t)$ (one writes $X_1$ and $\varphi_1$ in order to avoid any confusion with the quantities above)

$$x''(t) = X_1 e^{-\lambda t}\cos(\omega_1 t - \varphi_1) \tag{6.58}$$

The initial conditions $x(0) = 0$ and $\dot{x}(0) = 0$ allow one to calculate the unknowns $X_1$ and $\varphi_1$ and to find again the preceding result, given by (6.54) and (6.55).

### 6.6.2 Frequency response to a rectangular excitation

Calculate the response $x(t)$ of an oscillator to a rectangular excitation $f(t)$ by using the Fourier transform. Present the frequency spectra associated with $f(t)$ and $x(t)$.

Fig. 6.5 Rectangular exciting force.

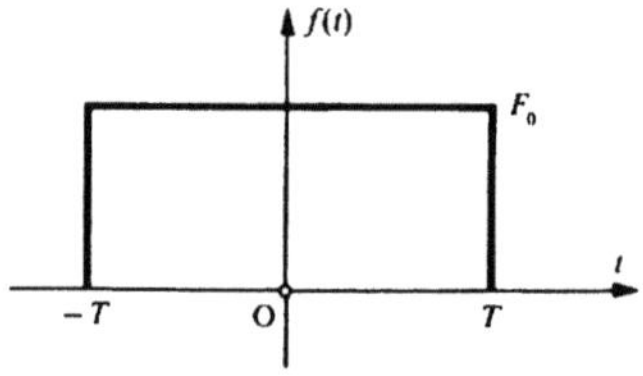


The function f(t) is defined by

$$f(t) = F_0 \quad \text{for} \quad -T < t < T$$
$$f(t) = 0 \quad \text{for} \quad t < -T \text{ and } t > T$$

The differential equation of the system is written

$$m \ddot{x} + c \dot{x} + k x = f(t)$$

By dividing the two sides by the stiffness k and by adopting the usual notation, this gives

$$\frac{1}{\omega_0^2} \ddot{x} + \frac{2 \eta}{\omega_0} \dot{x} + x = \frac{1}{k} f(t) \tag{6.59}$$

Taking the Fourier transform of each side, one can write

$$\left(1 - \left(\frac{\omega}{\omega_0}\right)^2 + 2 j \eta \frac{\omega}{\omega_0}\right) \underline{X}(\omega) = \frac{1}{k} \underline{F}(\omega) \tag{6.60}$$

or again

$$\underline{X}(\omega) = \frac{1}{1 - \left(\frac{\omega}{\omega_0}\right)^2 + 2 j \eta \frac{\omega}{\omega_0}} \frac{1}{k} \underline{F}(\omega) \tag{6.61}$$

One recognizes, on the right-hand side, the complex frequency response $\underline{H}(\omega)$ defined by (4.24). One clearly finds relation (6.52) again

$$\underline{X}(\omega) = \frac{1}{k}\,\underline{H}(\omega)\,\underline{F}(\omega) \tag{6.62}$$

The function f(t) having only two discontinuities and no infinite discontinuity, satisfies the condition of Dirichlet. The Fourier transform of f(t) is given by the expression

$$\underline{F}(\omega) = \int_{-\infty}^{\infty} f(t)\, e^{-j\omega t}\, dt = F_0 \int_{-T}^{T} e^{-j\omega t}\, dt$$

$$\underline{F}(\omega) = \frac{F_0}{j\,\omega}\,(e^{j\omega T} - e^{-j\omega T}) = 2\,F_0\,\frac{\sin \omega T}{\omega} \tag{6.63}$$

By using (6.62) and (6.63), the Fourier transform of the response is written

$$\underline{X}(\omega) = \frac{2F_0}{k}\,\frac{\sin \omega T}{\omega[1 - (\frac{\omega}{\omega_0})^2 + 2j\,\eta\,(\frac{\omega}{\omega_0})]} \tag{6.64}$$

We are going to determine the real and imaginary parts, then the modulus and the phase of $\underline{X}(\omega)$ . It gives successively

$$\mathrm{Re}[\underline{X}(\omega)] = \frac{2F_0}{k}\,\frac{\sin\omega t}{\omega}\,\frac{1 - (\frac{\omega}{\omega_0})^2}{[1 - (\frac{\omega}{\omega_0})^2]^2 + 4\eta 2(\frac{\omega}{\omega_0})^2} \tag{6.65}$$

$$\mathrm{Im}[\underline{X}(\omega)] = \frac{2F_0}{k}\,\frac{\sin\omega t}{\omega}\,\frac{-\,2\eta\;(\frac{\omega}{\omega_0})^2}{[1 - (\frac{\omega}{\omega_0})^2]^2 + 4\eta 2(\frac{\omega}{\omega_0})^2} \tag{6.66}$$

$$|\underline{X}(\omega)| = \frac{2F_0}{k}\left|\frac{\sin \omega T}{\omega}\right|\frac{1}{[(1 - (\frac{\omega}{\omega_0})^2)^2 + 4\,\eta^2\,(\frac{\omega}{\omega_0})^2]^{1/2}} \tag{6.67}$$

$$\psi(\omega) = \varphi(\omega) + \frac{\pi}{2}\,(\mathrm{sgn}(\omega) - \mathrm{sgn}(\sin\omega T)) \tag{6.68}$$

with

$$\mathrm{tg}\ \varphi(\omega) = \frac{-\,2\eta\,\frac{\omega}{\omega_0}}{1 - (\frac{\omega}{\omega_0})^2} \tag{6.69}$$

The spectrum associated with the exciting force $f(t)$ is shown in figure 6.6. Regarding the functions (6.65) to (6.68), they are illustrated for the damping factor $\eta = 0{,}25$, by the figures 6.7 to 6.10.

For the search for the inverse transform, it is advantageous to decompose (6.64) into simple elements

$$\underline{X}(\omega) = \frac{F_0}{k}\left[\frac{2\sin\omega T}{\omega} - \left(\frac{\lambda}{\omega_1} + j\right)\frac{\sin\omega T}{\lambda + j\omega - j\omega_1} + \left(\frac{\lambda}{\omega_1} - j\right)\frac{\sin\omega T}{\lambda + j\omega + j\omega_1}\right] \tag{6.70}$$

with

$$\lambda = \eta\omega_0 \qquad \text{and} \qquad \omega_1 = \omega_0\sqrt{1-\eta^2}$$

By expressing $\sin\omega T$ in its exponential form, the time response is written using (6.51)

$$X(t) = \frac{F_0}{k}\left[\frac{1}{2\pi}\int_{-\infty}^{\infty}\frac{e^{j\omega(t+T)} - e^{j\omega(t-T)}}{j\omega}\,d\omega - \frac{1}{4\pi}\left[1 - j\frac{\lambda}{\omega_1}\right]\int_{-\infty}^{\infty}\frac{e^{j\omega(t+T)} - e^{j\omega(t-T)}}{\lambda + j\omega - j\omega_1}\,d\omega - \frac{1}{4\pi}\left[1 + j\frac{\lambda}{\omega_1}\right]\int_{-\infty}^{\infty}\frac{e^{j\omega(t+T)} - e^{j\omega(t-T)}}{\lambda + j\omega + j\omega_1}\,d\omega\right] \tag{6.71}$$

In order to evaluate this expression, it is necessary to know the values of the three following integrals, which one can calculate using contour integrals in the complex plane

$$\int_{-\infty}^{\infty}\frac{e^{j\omega u}}{j\,\omega}\,d\omega = \begin{cases} -\pi & \text{if } u < 0 \\ \pi & \text{if } u > 0 \end{cases}$$

$$\int_{-\infty}^{\infty}\frac{e^{j\omega u}}{\lambda + j\omega - j\omega_1}\,d\omega = \begin{cases} 0 & \text{if } u < 0 \\ 2\,\pi\,e^{-\lambda u}e^{j\omega_1 u} & \text{if } u > 0 \end{cases} \tag{6.72}$$

$$\int_{-\infty}^{\infty}\frac{e^{j\omega u}}{\lambda + j\omega + j\omega_1}\,d\omega = \begin{cases} 0 & \text{if } u < 0 \\ 2\,\pi\,e^{-\lambda u}e^{-j\omega_1 u} & \text{if } u > 0 \end{cases}$$

The variable $u$ has the values

$$u_1 = t + T$$
$$u_2 = t - T$$

One must therefore consider the three intervals of time

$$t < -T \qquad u_1 < 0 \ , \ u_2 < 0$$
$$-T < t < T \qquad u_1 > 0 \ , \ u_2 < 0$$
$$T < t \qquad u_1 > 0 \ , \ u_2 > 0$$

One obtains successively

· $t < -T$

$$x(t) = \frac{F_0}{k} \left(\frac{1}{2\pi} [-\pi + \pi] - 0 - 0 \right) = 0 \tag{6.73}$$

· $-T < t < T$

$$x(t) = \frac{F_0}{k} \left(\frac{1}{2\pi} [\pi + \pi] - \frac{1}{2} (1 - j \frac{\lambda}{\omega_1}) e^{-\lambda(t+T)} e^{j\omega_1(t+T)} - \frac{1}{2} (1 + j \frac{\lambda}{\omega_1}) e^{-\lambda(t+T)} e^{-j\omega_1(t+T)}\right) \tag{6.74}$$

$$x(t) = \frac{F_0}{k} \left(1 - e^{-\lambda(t+T)} [\cos\omega_1(t+T) + \frac{\lambda}{\omega_1} \sin\omega_1(t+T)]\right)$$

· $T < t$

$$x(t) = \frac{F_0}{k} \left(\frac{1}{2\pi} (\pi - \pi) - \frac{1}{2} [1 - j\frac{\lambda}{\omega_1}][e^{-\lambda(t+T)} e^{j\omega_1(t+T)} - e^{-\lambda(t-T)} e^{j\omega_1(t-T)}] - \frac{1}{2} [1 + j\frac{\lambda}{\omega_1}][e^{-\lambda(t+T)} e^{-j\omega_1(t+T)} - e^{-\lambda(t-T)} e^{-j\omega_1(t-T)}]\right)$$

$$x(t) = \frac{F_0}{k} \left(e^{-\lambda(t-T)} [\cos\omega_1(t-T) + \frac{\lambda}{\omega_1} \sin\omega_1(t-T)] - e^{-\lambda(t+T)} [\cos\omega_1(t+T) + \frac{\lambda}{\omega_1} \sin\omega_1(t+T)]\right) \tag{6.75}$$

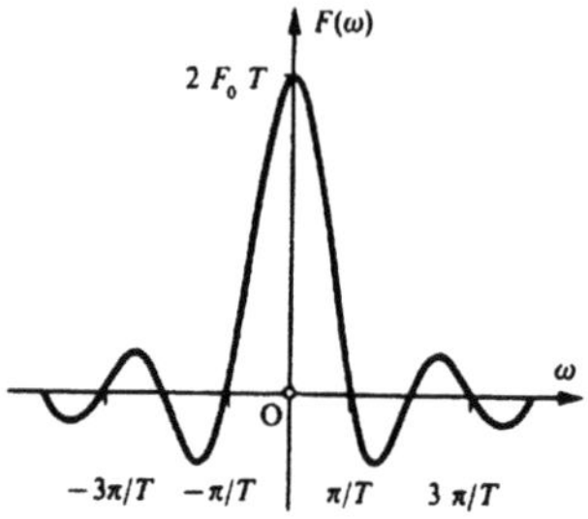


Fig. 6.6 Spectrum associated with the exciting force f(t)

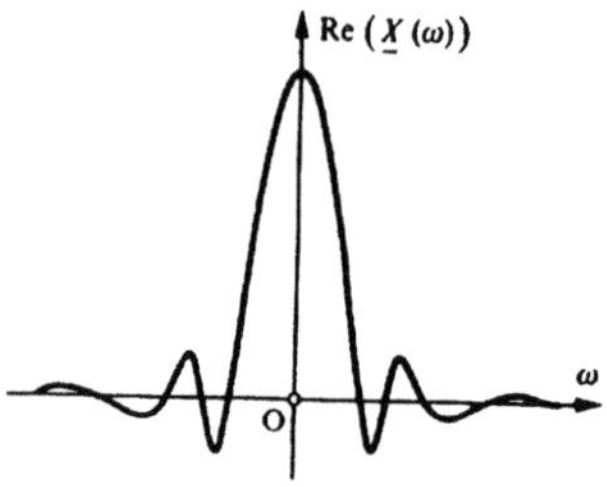


Fig. 6.7 Spectrum associated with the real part of the response $\underline{X}(\omega)$

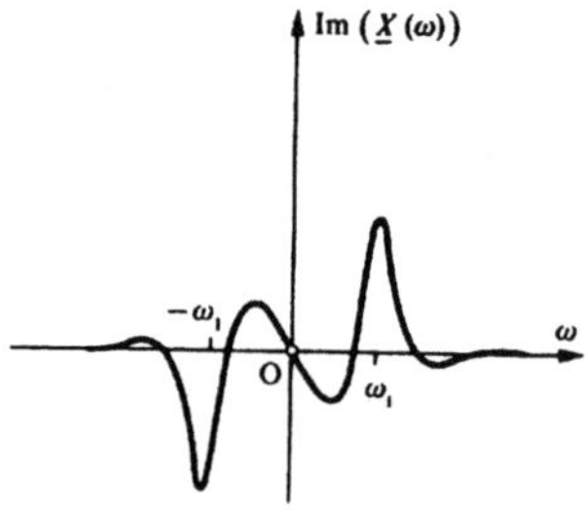


Fig. 6.8 Spectrum associated with the imaginary part of the response $\underline{X}(\omega)$

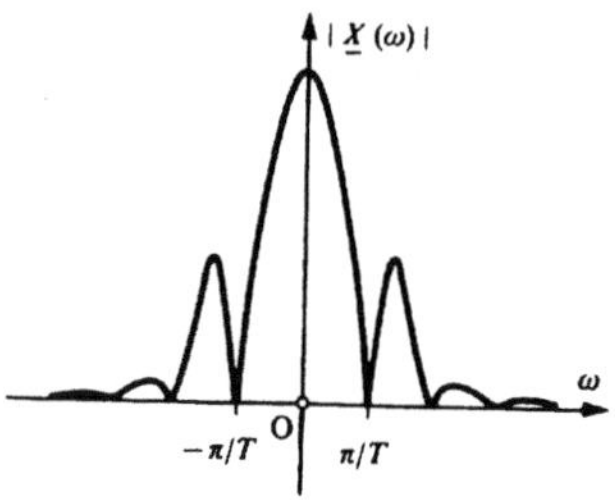


Fig. 6.9 Spectrum associated with the modulus $|X(\omega)|$ of the response x(t)

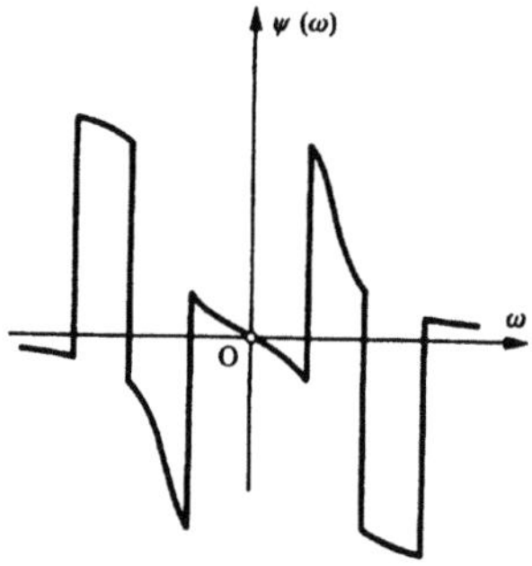


Fig. 6.10 Spectrum associated with the phase of the response x(t)

Figure 6.11 shows the shape of this response

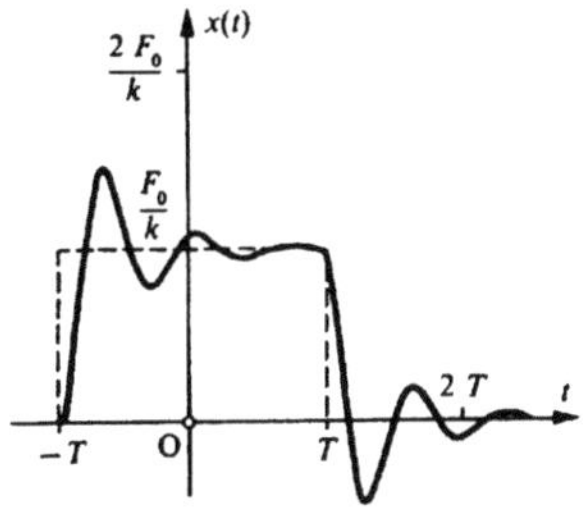


Fig. 6.11 Response x(t) of the dissipative oscillator

The response of the conservative system can be obtained by taking the limit of $x(t)$ when $\eta$ tends to zero. One obtains

$$\lim_{\eta \to 0} \lambda = 0 \qquad \lim_{\eta \to 0} \omega_1 = \omega_0$$

• $t < T$

$$x(t) = 0 \tag{6.76}$$

• $- T < t < t$

$$x(t) = \frac{F_0}{k} [1 - \cos \omega_0 (t+T)] \tag{6.77}$$

• $T < t$

$$x(t) = \frac{F_0}{k} [\cos\omega_0 (t-T) - \cos\omega_0 (t+T)] = \frac{F_0}{k} 2\sin\omega_0 T \sin\omega_0 t \tag{6.78}$$

Figure 6.12 shows this response in the case $T/T_0 = 1.43$ and $T_0 = 2\pi/\omega_0$

One notes that this response is not zero for $t > T$ if $T \neq n\pi/\omega_0 = nT_0/2$ for integer values of n. Since a response of the type $x(t) = \sin\omega_0 t$ does not satisfy the condition (6.48), the transition to the limit $\eta = 0$ must be made after integration. In fact

$$\lim_{\eta \to 0} \int_{-\infty}^{\infty} \frac{e^{j\omega u}}{\lambda + j\omega - j\omega_1} d\omega \neq \int_{-\infty}^{\infty} \lim_{\eta \to 0} \left(\frac{e^{j\omega u}}{\lambda + j\omega - j\omega_1}\right) d\omega \tag{6.79}$$

One sees therefore that the calculation of the response of a conservative system by means of the Fourier transform requires some specific precautions.

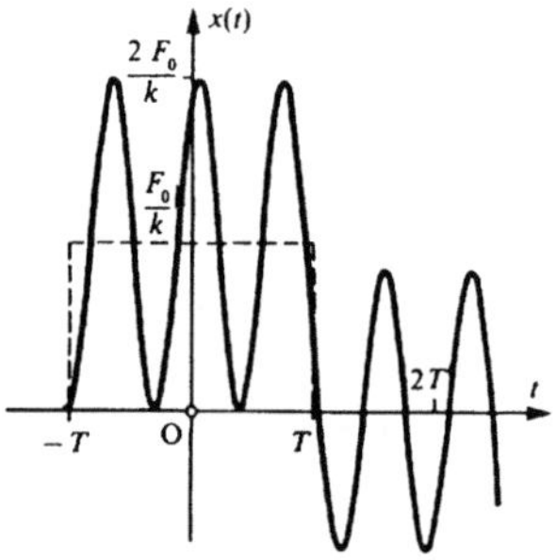


Fig. 6.12 Response x(t) of the conservative oscillator

Comments

- In contrast to the result for example 5.3.2. one notes here that the spectra associated with the exciting force and the response x(t) are continuous.

- The predominance of the fundamental, of angular frequency $\omega_1$ , is illustrated by the peaks close to $\pm\omega_1$ (figures 6.8 and 6.9). The amplitude of these peaks tends towards infinity when the damping tends to zero.

# CHAPTER 7 ELECTRICAL ANALOGUES

## 7.1 Generalities

As we said in the introduction, it is sometimes advantageous to search for the electrical system equivalent to a mechanical system.

The usefulness of the equivalent electrical system no longer lies in the possibility of constructing the system, as in the past, in order to measure the electrical quantities corresponding to the mechanical quantities, whose measurement was, at that time, relatively difficult. Instead, the modern methods of numerical analysis are more convenient and richer in possibilities. Rather the interest in an electric schematic lies in the ease of interpretation of the role played by the different elements of the system.

The elementary oscillator of mechanics consists of three discrete elements, a mass, a spring and a damper. It is the same for the elementary electrical oscillator which consists of a capacity, a self-inductance and a resistance.

If one wants to establish an analogy between the two oscillators, the dissipative elements, that is to say the damper and the resistance, are necessarily going to correspond. For the conservative elements, there would appear to be two possibilities :

· the mass corresponds to the self-inductance and the spring to the capacity; this is the **force-tension analogy**.

· the mass corresponds to the capacity and the spring to the self-inductance; this is the **force-current analogy**.

Only the second analogy will be considered in this chapter for it is preferable in nearly all practical problems. In fact, it is generally the external forces which are known, so it is more convenient to consider the sums of currents than the sums of voltages in the reasoning.

## 7.2 Force-current analogy

The force-current analogy, also called the mobility analogy, for the free state elementary oscillator gives rise to the correspondence shown in the electrical circuit diagram of figure 7.1.

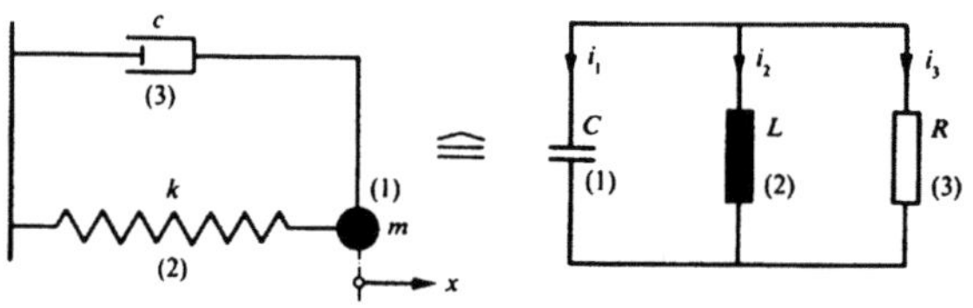


Fig. 7.1 Force-current analogy · Free state

The sum of the three currents is zero, as is the sum of the forces acting on the mass :

$$i_1 + i_2 + i_3 = 0 \tag{7.1}$$

On the other hand, the equality of the voltage drops gives

$$\frac{1}{C}\int i_1 \, dt = L \frac{di_2}{dt} = R \, i_3 \tag{7.2}$$

By eliminating $i_1$ and $i_3$

$$i_1 = L \, C \frac{d^2 i_2}{dt^2} \qquad\qquad i_3 = \frac{L}{R} \frac{di_2}{dt}$$

Equation (7.1) becomes

$$L \, C \frac{d^2 i_2}{dt^2} + i_2 + \frac{L}{R} \frac{di_2}{dt} = 0$$

Let us introduce the flux in the self-inductance $\Phi = L \, i_2$

$$C \, \ddot{\Phi} + \frac{\dot{\Phi}}{R} + \frac{\Phi}{L} = 0 \tag{7.3}$$

The two equivalent equations (7.1) and (7.3), compared to that of the mechanical oscillator

$$m \ddot{x} + c \dot{x} + k x = 0$$

lead to the correspondences below

Force-currents

| | |
|---|---|
| $m \ddot{x} \triangleq i_1$ | the force of inertia corresponds to the current in the capacity, |
| $k x \triangleq i_2$ | the elastic returning force corresponds to the current in the self-inductance, |
| $c \dot{x} \triangleq i_3$ | the force of viscous resistance corresponds to the current in the electrical resistance. |

Displacement-flux and their derivatives

| | |
|---|---|
| $x \triangleq \Phi$ | the displacement corresponds to the flux in the self-inductance, |
| $v = \dot{x} \triangleq \dot{\Phi} = u$ | the velocity (equal for the three mechanical elements) corresponds to the voltage (equal for the three electrical elements). |

Characteristic of the two systems

| | |
|---|---|
| $m \triangleq C$ | the mass corresponds to the capacity, |
| $k \triangleq \frac{1}{L}$ | the stiffness corresponds to the inverse of the self-inductance, |
| $c \triangleq \frac{1}{R}$ | the viscous mechanical resistance corresponds to the inverse of the electrical resistance. |

Energies

$\frac{m\ v^2}{2} \triangleq \frac{C\ u^2}{2}$ the kinetic energy corresponds to the electrostatic energy

$\frac{k\ x^2}{2} \triangleq \frac{\varphi^2}{2\ L}$ the potential energy corresponds to the electromagnetic energy

$c\ v^2 \triangleq \frac{u^2}{R}$ the power dissipated in the damper corresponds to the power dissipated in the electrical resistance

$= R\ i_3^2$

In the forced state, the equivalent circuit diagrams for the force-current analogy can be found readily by qualitative reasoning. We are going to consider two examples.

An harmonic force acting on a mass

The force corresponds to the current, the equivalent of a **force generator** $f = F \cos \omega t$ is a **current source** $i = I \cos \omega t$.

The force $f$ is divided into three parts :

- the first accelerates the mass; it corresponds to the current in the capacity;
- the second displaces the spring; it corresponds to the current in the self-inductance;
- the third displaces the damper, it corresponds to the current in the resistance.

One obtains in this way the circuit diagram of figure 7.2.

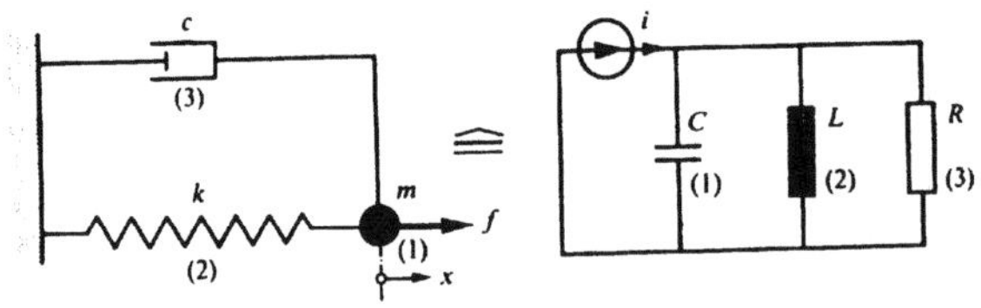


Fig. 7.2 Force-current analogy · A force acts on the mass

A displacememt $x = X \cos \omega t$ is imposed on the mass

An identical velocity $\dot{x} = -\omega X \sin \omega t = V \cos \omega t'$ is given to the three elements of the system. With velocity corresponding to voltage, the equivalent of a **velocity generator** is a **voltage generator** $u = U \cos \omega t'$ which supplies in parallel the three elements of the electrical circuit diagram shown in figure 7.3.

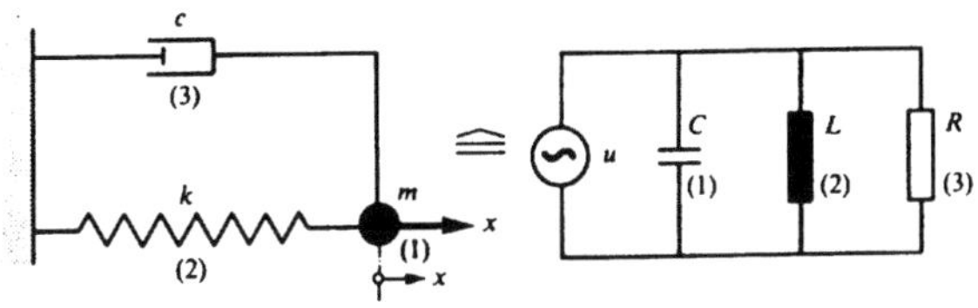


Fig. 7.3 Force-current analogy · A mass is displaced

## 7.3 Extension to systems with the several degrees of freedom · Circuits of forces

The reasoning made previously can be extended easily, in general, to systems with several degrees of freedom. A complete justification, by comparing differential equations, presents hardly any interest. It is in any case beyond the scope of this chapter, which is devoted to the electrical analogues of the elementary oscillator.

We limit ourselves to the example of figure 7.4. It concerns the mobility analogy of a system with three degrees of freedom being acted upon by an harmonic force.

In the mobility analogy, the correspondences are established as follows.

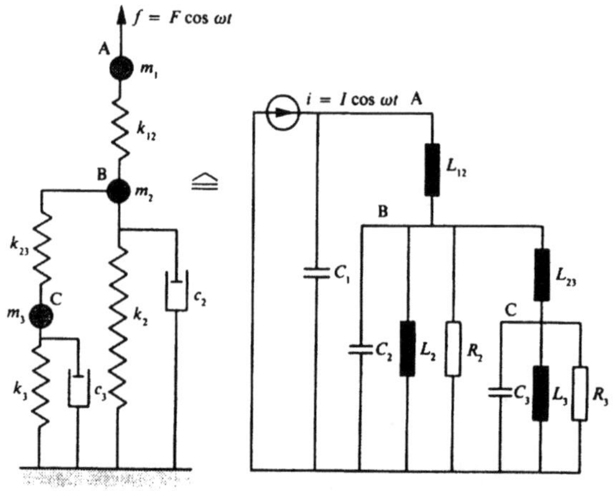


Fig. 7.4 Mobility analogy · System with three degrees of freedom

Level A 2 forces → 2 currents

inertial force on $m_1$ ≙ current in $C_1$

elastic force on $k_{12}$ ≙ current in $L_{12}$

Level B 4 forces → 4 currents

inertial force on $m_2$ ≙ current in $C_2$

elastic force on $k_2$ ≙ current in $L_2$

viscous force on $c_2$ ≙ current in $R_2$

elastic force on $k_{23}$ ≙ current in $L_{23}$

Level C 3 forces → 3 currents

inertial force on $m_3$ ≙ current in $C_3$

elastic force on $k_3$ ≙ current in $L_3$

viscous force on $c_3$ ≙ current in $R_3$

Comparison of the circuit diagrams of figure 7.4 leads to the following comments :

- The condition that the sums of currents are zero at the nodes (equivalent to the sums of the forces), as well as the possibilities for resonances can both be seen more clearly in the electrical circuit diagram.

- On the other hand, if one asks how many degrees of freedom the system has, one immediately sees that the answer is 3 for the mechanical circuit diagram (position of 3 masses, for example), whereas that demands a little more reflection for the electrical circuit diagram (voltage at the level A, B, C, for example).

It is possible to represent the mechanical circuit diagram by an electrical circuit diagram for the identical equivalent circuit by proceeding in the way illustrated in figure 7.5 for the case of an oscillator excited by an harmonic force $F \cos \omega t$ . This figure includes the usual mechanical schematic (a), the modified mechanical schematic (circuit of forces) with the graphic symbol for an harmonic force generator (b), and the analogous electrical circuit diagram (c).

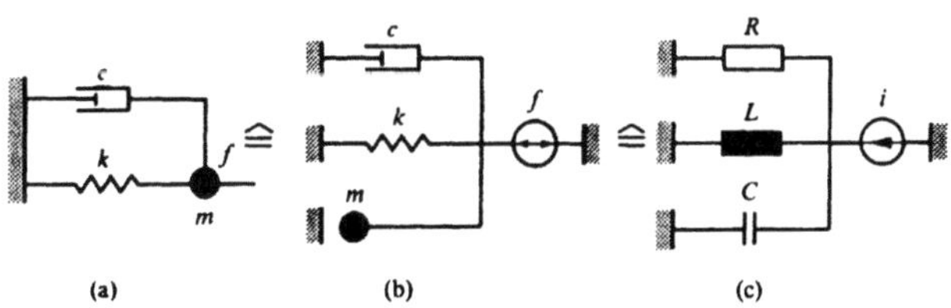


Fig. 7.5 Mobility analogy · Circuit of forces

By adopting this technique, the two schematics of figure 7.4 take the form shown in figure 7.6.

Consequently the mobility analogy leads to the concept of **circuits of forces**, completely equivalent to the circuits of electric currents. With a bit of practice, the representation of the electrical circuit diagram

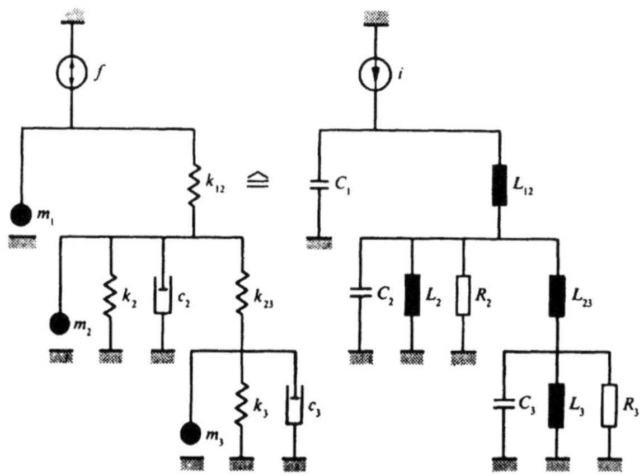


Fig. 7.6 Mobility analogy · Circuit of forces · System with three degrees of freedom

corresponding to the mechanical system is no longer necessary. One directly represents the circuits of forces and one calculates or measures the mechanical impedances of the system studied. One ends up in this way with the **method of mechanical impedances** which has acquired some importance in recent years, mainly at the experimental level. In particular, this method is advantageous for the study of complex systems, which can be broken down into several sub-systems whose analysis is easier. Bringing the results together then enables one to describe the behaviour of the complete system, for certain well defined circumstances.

However it would be dangerous to think that all the methods of analysis and the results of the theory of electrical circuits can be transposed directly to the circuits of forces. In fact, the measurement of mechanical impedances is more arduous than that for electrical impedances and, above all, the possibilities for correctly isolating one sub-system from another are generally much more limited.

# CHAPTER 8 SYSTEMS WITH TWO DEGREES OF FREEDOM

## 8.1 Generalities · Concept of coupling

A mechanical system possesses two degrees of freedom, when its configuration can be describe by means of two functions of time $x_1(t)$ and $x_2(t)$ , which are called generalized coordinates in the sense of Lagrangian mechanics. These can represent lengths (m), angles (radians), volumes ($m^3$), etc. If the system is linear and its characteristics are constant, then, in the most general case, its behaviour in the free state is governed by the following two differential equations

$$\begin{cases} m_{11}\ddot{x}_1 + c_{11}\dot{x}_1 + k_{11}x_1 + m_{12}\ddot{x}_2 + c_{12}\dot{x}_2 + k_{12}x_2 = 0 \\ m_{22}\ddot{x}_2 + c_{22}\dot{x}_2 + k_{22}x_2 + m_{21}\ddot{x}_1 + c_{21}\dot{x}_1 + k_{21}x_1 = 0 \end{cases} \tag{8.1}$$

| Natural terms | Inertial coupling terms | Resistive coupling terms | Elastic coupling terms |
|---|---|---|---|

Each equation includes three natural terms, as well as three coupling terms representing the action of $x_2$ on $x_1$ , or of $x_1$ on $x_2$ respectively. All these terms have the nature of generalized forces, that is to say of actual forces (N), of moments (N.m), of pressures ($N/m^2$), etc. Consequently, the equations signify that the two sums of six forces are zero. The terms $m_{12}\ddot{x}_2$ and $m_{21}\ddot{x}_1$ represent the forces of **inertial coupling** (or mass coupling). The terms $c_{12}\dot{x}_2$ and $c_{21}\dot{x}_1$ represent the forces of **resistive coupling** of a viscous nature (linear resistive). Finally, the terms $k_{12}x_2$ and $k_{21}x_1$ correspond to the forces of **elastic coupling,** the most frequent in practice.

A complete study of systems with two degrees of freedom, such as we have made for the elementary oscillator, will not be made here. On the one hand, it would be extremely tedious because of the number of existing parameters (eleven instead of two for the elementary oscillator), on the other hand the essential properties of the solutions will be established later, in a general way, for a linear oscillating system comprising $n$ degrees of freedom.

In this chapter, we shall limit ourselves to studying the free state of the conservative system, then to studying elastic coupling in a detailed way. The following chapter will tackle an example of the forced state, the Frahm oscillator.

Let us return to equations (8.1) put in matrix form

$$\begin{bmatrix} m_{11} & m_{12} \\ m_{21} & m_{22} \end{bmatrix} \begin{Bmatrix} \ddot{x}_1 \\ \ddot{x}_2 \end{Bmatrix} + \begin{bmatrix} c_{11} & c_{12} \\ c_{21} & c_{22} \end{bmatrix} \begin{Bmatrix} \dot{x}_1 \\ \dot{x}_2 \end{Bmatrix} + \begin{bmatrix} k_{11} & k_{12} \\ k_{21} & k_{12} \end{bmatrix} \begin{Bmatrix} x_1 \\ x_2 \end{Bmatrix} = \begin{Bmatrix} 0 \\ 0 \end{Bmatrix} \quad (8.2)$$

or, by compressing the writing

$$[M]\,\ddot{\vec{x}} + [C]\,\dot{\vec{x}} + [K]\,\vec{x} = \vec{0} \quad (8.3)$$

One adopts the following definitions,

$\vec{x}$ vector of displacements,

$\dot{\vec{x}}$ vector of velocities,

$\ddot{\vec{x}}$ vector of accelerations,

[M] mass matrix (or matrix of coefficients of inertia),

[C] damping matrix (or loss matrix),

[K] stiffness matrix (or rigidity matrix).

These definitions will be retained in the study of the generalized oscillator.

As an example, let us find the terms of the matrices for the standard schematic of figure 8.1, corresponding to a dissipative oscillator with two degrees of freedom, without inertial coupling.

Newton's equations for the system are written

$$\begin{cases} m_1\,\ddot{x}_1 = -\,k_1\,x_1 - k_3(x_1 - x_2) - c_1\,\dot{x}_1 - c_3(\dot{x}_1 - \dot{x}_2) \\ m_2\,\ddot{x}_2 = -\,k_2\,x_2 - k_3(x_2 - x_1) - c_2\,\dot{x}_2 - c_3(\dot{x}_2 - \dot{x}_1) \end{cases}$$

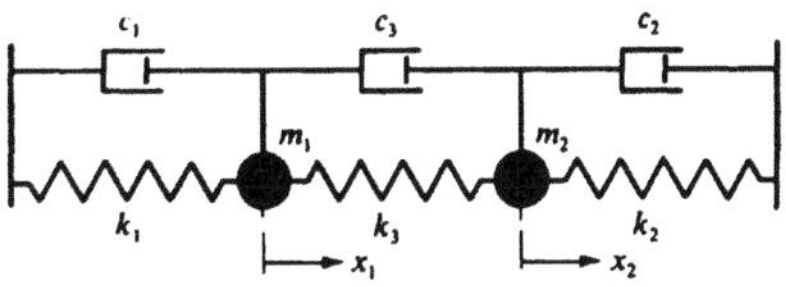


Fig. 8.1 Standard diagram for an oscillator with two degrees of freedom

or again

$$\begin{cases} m_1 \ddot{x}_1 + (c_1 + c_3)\dot{x}_1 - c_3 \dot{x}_2 + (k_1 + k_3)x_1 - k_3 x_2 = 0 \\ m_2 \ddot{x}_2 - c_3 \dot{x}_1 + (c_2 + c_3)\dot{x}_2 - k_3 x_1 + (k_2 + k_3)x_2 = 0 \end{cases} \tag{8.4}$$

Comparison between relations (8.2) and (8.4) gives

$$[M] = \begin{bmatrix} m_1 & 0 \\ 0 & m_2 \end{bmatrix} \qquad [C] = \begin{bmatrix} c_1 + c_3 & - c_3 \\ - c_3 & c_2 + c_3 \end{bmatrix} \qquad [K] = \begin{bmatrix} k_1 + k_3 & - k_3 \\ - k_3 & k_2 + k_3 \end{bmatrix} \tag{8.5}$$

The three matrices are symmetrical. We shall show later that it is always so for a discrete linear oscillator.

## 8.2 Free state and natural modes of the conservative system

When the resistances of figure 8.1 are zero, equations (8.4) take the simple form

$$\begin{cases} m_1 \ddot{x}_1 + (k_1 + k_3)x_1 - k_3 x_2 = 0 \\ m_2 \ddot{x}_2 - k_3 x_1 + (k_2 + k_3)x_2 = 0 \end{cases} \tag{8.6}$$

These are linear differential equations for which one seeks exponential solutions of the form

$$x_1 = A_1 e^{pt} \qquad x_2 = A_2 e^{pt}$$

One thus obtains the algebraic conditions

$$\begin{cases} (m_1 p^2 + k_1 + k_3)A_1 - k_3 A_2 = 0 \\ - k_3 A_1 + (m_2 p^2 + k_2 + k_3)A_2 = 0 \end{cases} \qquad (8.7)$$

The solutions of this homogeneous system are not all zero (the zero solutions correspond to static equilibrium) only if its determinant is zero

$$\begin{vmatrix} m_1 p^2 + k_1 + k_3 & - k_3 \\ - k_3 & m_2 p^2 + k_2 + k_3 \end{vmatrix} = 0 \qquad (8.8)$$

that is to say, on expansion

$$p^4 + p^2\left(\frac{k_1 + k_3}{m_1} + \frac{k_2 + k_3}{m_2}\right) + \frac{k_1 k_2 + k_2 k_3 + k_3 k_1}{m_1 m_2} = 0 \qquad (8.9)$$

This equation, known as **the characteristic equation** or **the frequency equation**, allows four purely imaginary solutions, combined in pairs, $\pm j\omega_1$ and $\pm j\omega_2$ .

Indeed, the general form of equation (8.9) is :

$$p^4 + B p^2 + C = 0 \quad \Rightarrow \quad p^2 = \frac{1}{2}\left(- B \pm \sqrt{B^2 - 4 C}\right)$$

The quantities $B$ and $C$ being essentially positive, the two solutions $p^2$ are negative if the root is real, in other words, if the difference $B^2 - 4 C$ is positive

$$B^2 - 4 C = \left(\frac{k_1 + k_3}{m_1} + \frac{k_2 + k_3}{m_2}\right)^2 - 4 \frac{(k_1 + k_3)(k_2 + k_3) - k_3^2}{m_1 m_2}$$

$$= \left(\frac{k_1 + k_3}{m_1} - \frac{k_2 + k_3}{m_2}\right)^2 + 4 \frac{k_3^2}{m_1 m_2} > 0$$

One has thus effectively

$$p_1^2 < 0 \quad \Rightarrow \quad p_1 = \pm j \omega_1 \qquad\qquad p_2^2 < 0 \quad \Rightarrow \quad p_2 = \pm j \omega_2$$

Equations (8.7) being homogeneous, the amplitudes $A_i$ are only defined to a scale factor. One can then only determine the ratio $\beta_2$ between the amplitudes $A_2$ and $A_1$

$$\beta_2 = \frac{A_2}{A_1} = \frac{m_1\ p^2 + k_1 + k_3}{k_3} = \frac{k_3}{m_2\ p^2 + k_2 + k_3} \tag{8.10}$$

This ratio is a real number which takes two distinct values as a function of the solutions, $\beta_{21}$ for $\pm\ j\omega_1$ and $\beta_{22}$ for $\pm\ j\omega_2$. One obtains :

$$\left\{\begin{array}{l} \beta_{21} = \dfrac{k_1 + k_3 - m_1\ \omega_1^2}{k_3} = \dfrac{k_3}{k_2 + k_3 - m_2\ \omega_1^2} \\[2ex] \beta_{22} = \dfrac{k_1 + k_3 - m_1\ \omega_2^2}{k_3} = \dfrac{k_3}{k_2 + k_3 - m_2\ \omega_2^2} \end{array}\right. \tag{8.11}$$

Let $A_{11}$, $A_{12}$ and $A_{21}$, $A_{22}$ be the values of $A_1$ and $A_2$ when $\beta_2$ has the values $\beta_{21}$ and $\beta_{22}$ respectively

$$\beta_{21} = \frac{A_{21}}{A_{11}} \qquad\qquad \beta_{22} = \frac{A_{22}}{A_{12}}$$

The system of equations (8.6) consequently has the following specific solutions

for $x_1$ $\quad A_{11}\ e^{j\omega_1 t}\ ;\ A_{11}\ e^{-j\omega_1 t}\ ;\ A_{12}\ e^{j\omega_2 t}\ ;\ A_{12}\ e^{-j\omega_2 t}$

for $x_2$ $\quad A_{21}\ e^{j\omega_1 t}\ ;\ A_{21}\ e^{-j\omega_1 t}\ ;\ A_{22}\ e^{j\omega_2 t}\ ;\ A_{22}\ e^{-j\omega_2 t}$

The general solutions are obtained by linear combinations of the specific solutions

$$\left\{\begin{array}{l} x_1 = A_{11}(C_1\ e^{j\omega_1 t} + D_1\ e^{-j\omega_1 t}) + A_{12}\ (C_2\ e^{j\omega_2 t} + D_2\ e^{-j\omega_2 t}) \\[1ex] x_2 = A_{21}\ (C_1\ e^{j\omega_1 t} + D_1\ e^{-j\omega_1 t}) + A_{22}\ (C_2\ e^{j\omega_2 t} + D_2\ e^{-j\omega_2 t}) \end{array}\right. \tag{8.12}$$

The constants $C_1$ , $D_1$ , $C_2$ and $D_2$ are arbitrary; moreover, since the $A_{ij}$ are only defined to a scale factor, the solutions can be put in the form

$$\begin{cases} x_1 = X_1 \cos(\omega_1 t-\varphi_1) + X_2 \cos(\omega_2 t-\varphi_2) \\ x_2 = \beta_{21} X_1 \cos(\omega_1 t-\varphi_1) + \beta_{22} X_2 \cos(\omega_2 t-\varphi_2) \end{cases} \quad (8.13)$$

1st mode 2nd mode

or in the matrix form

$$\vec{x} = \vec{\beta}_1 X_1 \cos(\omega_1 t-\varphi_1) + \vec{\beta}_2 X_2 \cos(\omega_2 t-\varphi_2) \quad (8.14)$$

with

$$\vec{\beta}_1 = \begin{Bmatrix} 1 \\ \beta_{21} \end{Bmatrix} \qquad \vec{\beta}_2 = \begin{Bmatrix} 1 \\ \beta_{22} \end{Bmatrix} \quad (8.15)$$

The above equations include 4 constants of integration, $X_1$ , $X_2$ , $\varphi_1$ , $\varphi_2$ , which are functions of the initial conditions. They introduce the concept of **natural modes,** to which we will return in detail in the study of the generalized oscillator. A **natural mode** is the **motion of a system** about **a** natural **angular frequency** (or at a natural frequency which comes to the same thing).

We have assumed that the system examined does not include resistances. The natural angular frequencies $\omega_1$ and $\omega_2$ are thus equivalent to the angular frequency $\omega_0$ of the conservative elementary oscillator studied in section 3.1. It would have been more consistent to have adopted the notation $\omega_{01}$ and $\omega_{02}$ instead of $\omega_1$ and $\omega_2$. However, as no confusion is possible in this chapter, we prefer the shorter notation.

A system with two degrees of freedom thus possesses two natural modes. By choosing the initial conditions in a way that $X_2 = 0$ , the system oscillates according to the first mode only. It oscillates according to the second mode if $X_1 = 0$ . In the general case, the two modes exist together, but do not have mutual influence, in other words there is no exchange of energy from one to the other. This important property, called **orthogonality of the natural modes,** will be established later. It expresses the linear independance of the natural vectors $\vec{\beta}$ and is translated here by the two relations

$$\vec{\beta}_1^T \,[M]\, \vec{\beta}_2 = 0 \qquad\qquad \vec{\beta}_1^T \,[K]\, \vec{\beta}_2 = 0 \tag{8.16}$$

being, after development

$$\begin{cases} m_1 + m_2\, \beta_{21}\, \beta_{22} = 0 \\ k_1 + k_2\, \beta_{21}\, \beta_{22} + k_3(1-\beta_{21})(1-\beta_{22}) = 0 \end{cases} \tag{8.17}$$

The role of the natural modes appears clearly when the system has a geometric symmetry (figure 8.2). In this case

$$m_1 = m_2 = m \qquad\qquad k_1 = k_2 = k$$

and the characteristic equation (8.9) becomes

$$p^4 + p^2\, \frac{2(k + k_3)}{m} + \frac{k^2 + 2\,k\,k_3}{m^2} = 0 \tag{8.18}$$

It has the solutions

$$p^2 = -\,\frac{k + k_3}{m} \pm \frac{k_3}{m} \tag{8.19}$$

The natural angular frequencies are thus :

$$\omega_1^2 = \frac{k}{m} \qquad\qquad \omega_2^2 = \frac{k + 2\,k_3}{m} \tag{8.20}$$

The ratios $\beta_{21}$ and $\beta_{22}$ of relation (8.11) then take the values +1 and - 1 . In fact

$$\beta_{21} = \frac{k + k_3 - m\,\omega_1^2}{k_3} = +\,1 \qquad\qquad \beta_{22} = \frac{k + k_3 - m\,\omega_2^2}{k_3} = -\,1 \tag{8.21}$$

The first mode corresponds to identical oscillations of the two masses

$$x_1 = x_2 = X_1 \cos(\omega_1 t - \varphi_1) \qquad (8.22)$$

whereas the second mode corresponds to the oscillations being 180° out of phase

$$\begin{cases} x_1 = X_2 \cos(\omega_2 t - \varphi_2) \\ x_2 = - X_2 \cos(\omega_2 t - \varphi_2) \end{cases} \qquad (8.23)$$

Three examples of symmetrical oscillators with two degrees of freedom are shown in figures 8.2 to 8.4.

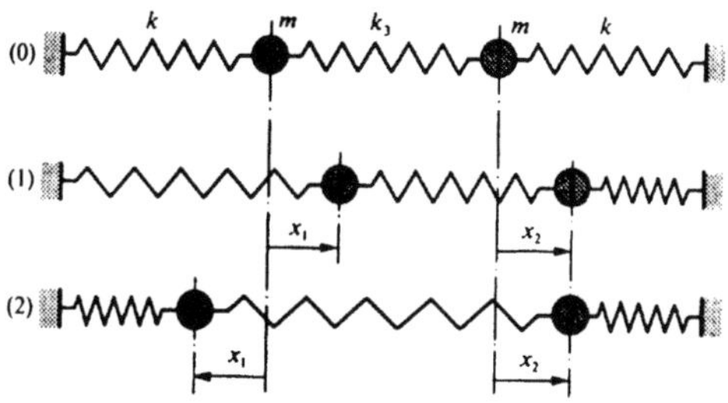


Fig. 8.2 Natural modes of a symmetrical oscillator with two degrees of freedom. Reference system
(0) system in static equilibrium
(1) 1st mode : identical displacements of the 2 masses $x_2 = x_1$
(2) 2nd mode : opposed displacement of the 2 masses $x_2 = - x_1$

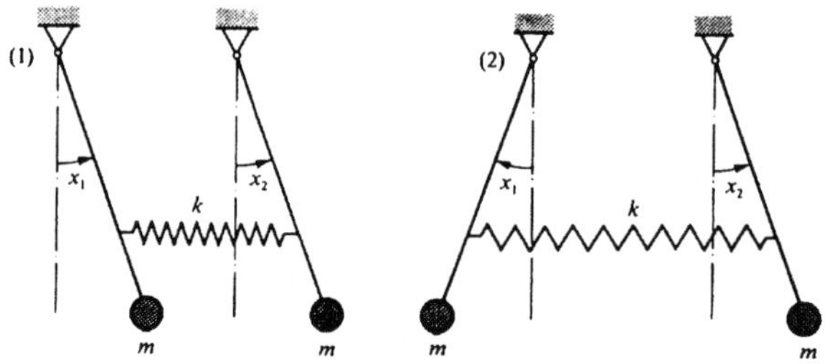


Fig. 8.3 Natural modes of a symmetrical double pendulum
(1) 1st mode : $x_2 = x_1$
(2) 2nd mode : $x_2 = \; x_1$

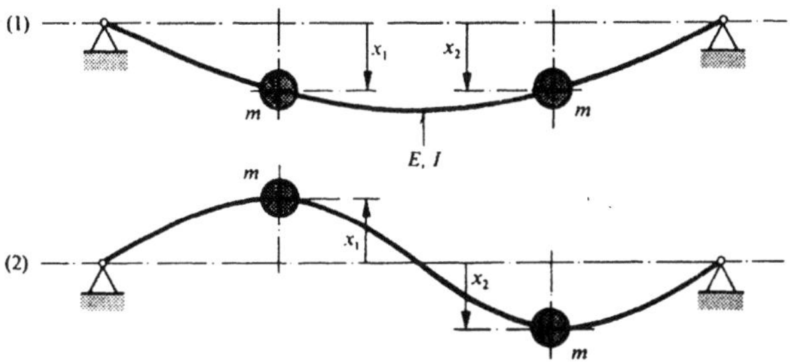


Fig. 8.4 Natural modes of the system consisting of two equal masses, symmetrically placed on a massless beam, oscillating by bending in a plane
(1) 1st mode : $x_2 = x_1$
(2) 2nd mode : $x_2 = - x_1$

One can make several comments concerning these examples :

- The static configuration corresponding to a natural mode is called a **mode shape** of the system. This concept will be generalized in what follows. Thus, the first and second mode shapes are shown in each of the figures.

- For the reference system and for the double pendulum, the elastic coupling does not participate in the first mode because of the symmetry, whereas it does in the second mode. This statement cannot be applied directly to the last example of the bending of a beam. However, one should note that the first mode shape has a single curved portion whereas the second has two, separated by an inflexion point.

- For an equal value of the displacement $x_1$ (for example the normalized value $|x_1| = 1$ ), the potential energy of the first mode is less than that of the second mode. It consists of the energy of deformation in the three examples together with, in the case of the double pendulum, the potential energy of position of the masses.

## 8.3 Study of elastic coupling

Let us return to the characteristic equation (8.8) by replacing p by $j\omega$, then by dividing the first line by $m_1$ and the second by $m_2$

$$\begin{vmatrix} \frac{k_1 + k_3}{m_1} - \omega^2 & -\frac{k_3}{m_1} \\ -\frac{k_3}{m_2} & \frac{k_2 + k_3}{m_2} - \omega^2 \end{vmatrix} = 0$$

The first term is a function of $\omega^2$ and one can write it, after expansion

$$f(\omega^2) = (\frac{k_1 + k_3}{m_1} - \omega^2)(\frac{k_2 + k_3}{m_2} - \omega^2) - \frac{k_3^2}{m_1 m_2} = 0 \qquad (8.24)$$

One introduces the **angular frequencies for zero coupling,** defined as follows (figure 8.5),

$\Omega_1^2 = \frac{k_1 + k_3}{m_1}$ angular frequency of $m_1$ when $m_2$ is blocked

$\Omega_2^2 = \frac{k_2 + k_3}{m_2}$ angular frequency of $m_2$ when $m_1$ is blocked

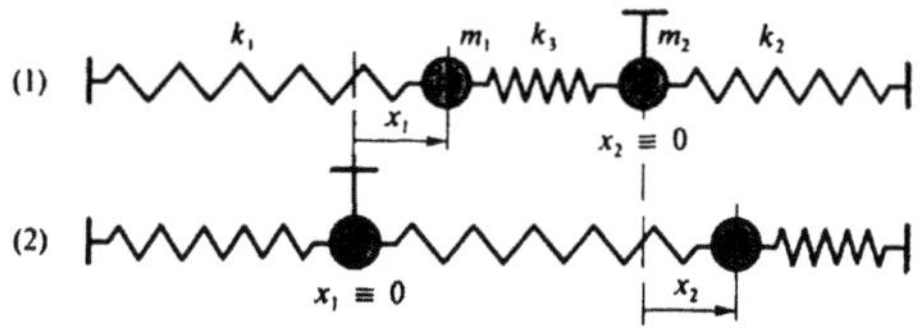


Fig. 8.5 Zero coupling in the reference system
(1) $m_2$ is coupled to zero
(2) $m_1$ is coupled to zero

The term characterizing the elastic coupling is designated by $\Omega_{12}$

$$\Omega_{12}^4 = \frac{k_3^2}{m_1\, m_2} \tag{8.25}$$

so equation (8.24) becomes

$$f(\omega^2) = (\Omega_1^2 - \omega^2)\,(\Omega_2^2 - \omega^2) - \Omega_{12}^4 = 0 \tag{8.26}$$

The function $f(\omega^2)$ is a parabola (figure 8.6) cutting the horizontal axis at the points $\omega_1^2$ and $\omega_2^2$ corresponding to the two natural angular frequencies of the system. The intersection of the parabola with the horizontal $f(\omega^2) = -\Omega_{12}^4$ gives the two angular frequencies for zero coupling.

These geometrical considerations lead to the following inequalities, which it is easy to establish algebraically

$$\omega_1^2 < \Omega_1^2 < \Omega_2^2 < \omega_2^2 \tag{8.27}$$

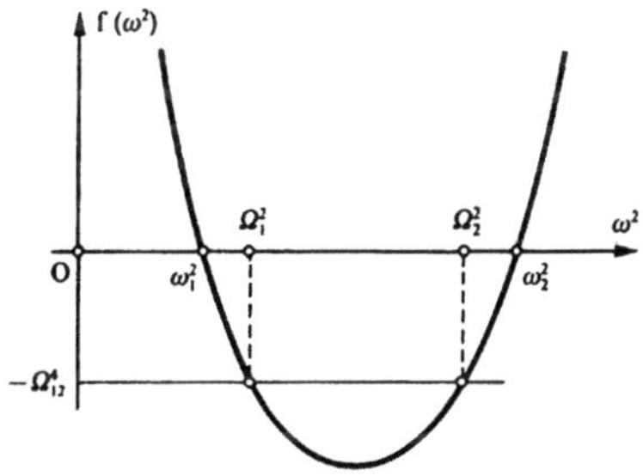


Fig. 8.6 Relative values of the natural angular frequencies and of the angular frequencies at zero coupling

Another geometrical representation can be proposed; it is inspired by Mohr's circles used for the study of the state of stress. In order to do that, let us calculate the roots of equation (8.26)

$$\omega^4 - \omega^2(\Omega_1^2 + \Omega_2^2) + \Omega_1^2\,\Omega_2^2 - \Omega_{12}^4 = 0$$

It gives thus

$$\omega^2 = \frac{1}{2}(\Omega_1^2 + \Omega_2^2) \pm \sqrt{\left(\frac{\Omega_2^2 - \Omega_1^2}{2}\right)^2 + \Omega_{12}^4} = \frac{1}{2}(\Omega_1^2 + \Omega_2^2) \pm R \qquad (8.28)$$

Fig. 8.7 Study of elastic coupling; circle of the angular frequencies

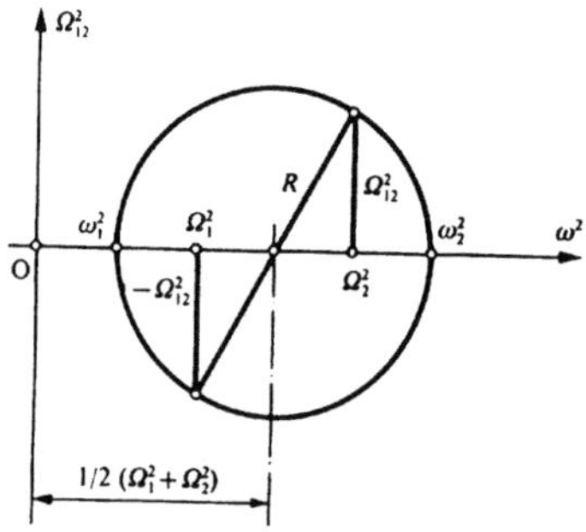


The natural frequencies $\omega_1^2$ and $\omega_2^2$ correspond to the ends of the horizontal diameter of a circle in the plane $\omega^2$, $\Omega_{12}^2$ (figure 8.7).

This presentation enables one to be aware immediately of the influence of the coupling on the gap between the natural angular frequencies.

When the coupling term $\Omega_{12}^2 = k_3/\sqrt{m_1\, m_2}$ is very large, the natural frequencies are cleanly differentiated from the angular frequencies $\Omega_1^2$ and $\Omega_2^2$. It is easy to show that they tend respectively towards the values

$$\omega_1^2 = \frac{k_1 + k_2}{m_1 + m_2} \qquad \omega_2^2 = k_3\,\frac{m_1 + m_2}{m_1\, m_2} \qquad (8.29)$$

If the stiffness $k_3$ tends to infinity, the system degenerates into an elementary oscillator of mass $m = m_1 + m_2$, of stiffness $k = k_1 + k_2$, and therefore of natural angular frequency $\omega_1^2$ given above. The circle of angular frequencies, in the domain of finite frequencies, degenerates into a vertical straight line at the abscissa $\omega_1^2$.

On the other hand, if the coupling tends towards zero, one has

$\omega_1^2 \to \Omega_1^2 \to \frac{k_1}{m_1}$ natural angular frequency of the mass $m_1$ with only one spring of stiffness $k_1$

$\omega_2^2 \to \Omega_2^2 \to \frac{k_2}{m_2}$ natural angular frequency of the mass $m_2$ with only one spring of stiffness $k_2$

Put in another way, if $k_3 \to 0$ , the system is transformed into two separate elementary systems.

## 8.4 Examples of oscillators with two degrees of freedom

### 8.4.1 Natural frequencies of a service lift

The service lift shown in figure 8.8 consists of a motor-gear-reduction group, a very long shaft because of the conditions of installation, a drum, a cable and a load.

By neglecting the masses of the shaft and of the cable, and assuming that the drum is rigid and that the cable always stays taut, calculate the natural frequencies and the angular frequencies for zero coupling in case of locking of the bearing $P_1$ .

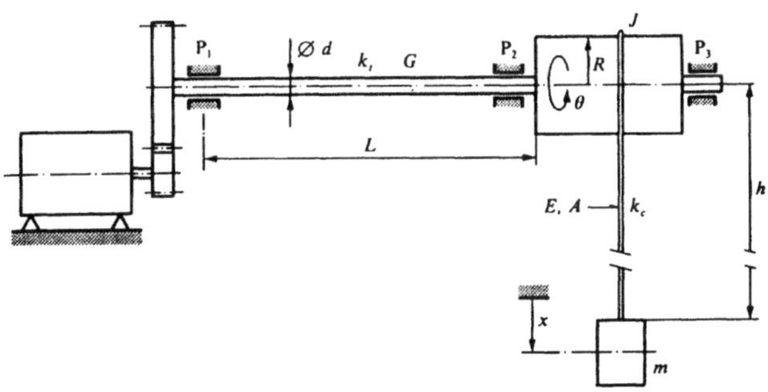


Fig. 8.8 Diagram of a service lift

<u>Data for the problem</u>

| | | |
|---|---|---|
| Shaft | diameter | $d = 50 \text{ mm} = 5\cdot 10^{-2} \text{ m}$ |
| | length | $L = 3 \text{ m}$ |
| | shear modulus | $G = 0.8\cdot 10^{11} \text{ N/m}^2$ |
| Drum | radius | $R = 200 \text{ mm} = 0{,}2 \text{ m}$ |
| | moment of inertia | $J = 5 \text{ kg}\cdot\text{m}^2$ |
| Cable | cross-section | $A = 1 \text{ cm}^2 = 10^{-4} \text{ m}^2$ |
| | length at the time of locking | $h = 15 \text{ m}$ |
| | modulus of elasticity | $E = 2.0\cdot 10^{11} \text{ N/m}^2$ |
| Load | mass | $m = 300 \text{ kg}$ |

By assuming the bearing $P_1$ to be locked and by taking the positions of static equilibrium as origins, the oscillations of the system can be described by the angle $\theta$ of rotation of the drum with respect to the bearing $P_1$ and the vertical displacement $x$ of the load.

One defines the quantities

$$k_t = \frac{G\,I_p}{L} \quad \text{torsion stiffness of the shaft} \quad \left(I_p = \frac{\pi d^4}{32}\right),$$

$$k_c = \frac{E\,A}{h} \quad \text{elongation stiffness of the cable}$$

The equations of motion are written :

$$\begin{cases} J\,\ddot{\theta} = -\,k_t\,\theta - R\,k_c(R\theta - x) \\ m\,\ddot{x} = -\,k_c(x - R\theta) \end{cases}$$

so that

$$\begin{cases} J\,\ddot{\theta} + (k_t + R^2\,k_c)\theta - R\,k_c\,x = 0 \\ m\,\ddot{x} + k_c\,x - R\,k_c\,\theta = 0 \end{cases} \tag{8.30}$$

By seeking solutions of the form

$$\theta = A_1\, e^{pt} = A_1\, e^{j\omega t} \qquad x = A_2\, e^{pt} = A_2\, e^{j\omega t}$$

one obtains the characteristic equation

$$\begin{vmatrix} (k_t + R^2\, k_c - J\, \omega^2) & - R\, k_c \\ - R\, k_c & (k_c - m\, \omega^2) \end{vmatrix} = 0$$

whence, after expansion

$$(\frac{k_c}{m} - \omega^2)\, (\frac{k_t + R^2\, k_c}{J} - \omega^2) - \frac{R^2\, k_c^2}{J\, m} = 0 \qquad (8.31)$$

It is convenient to introduce the angular frequencies for zero coupling and the coupling term defined in section 8.3

$$\Omega_1^2 = \frac{k_c}{m} \qquad \Omega_2^2 = \frac{k_t + R^2\, k_c}{J} \qquad \Omega_{12}^4 = \frac{R^2\, k_c^2}{J\, m} \qquad (8.32)$$

Equation (8.31) becomes in this way

$$(\Omega_1^2 - \omega^2)\, (\Omega_2^2 - \omega^2) - \Omega_{12}^4 = 0$$

One determines the natural angular frequencies from it using (8.28)

$$\omega^2 = \frac{1}{2}\, (\Omega_1^2 + \Omega_2^2) \pm \sqrt{(\frac{\Omega_2^2 - \Omega_1^2}{2})^2 + \Omega_{12}^4} \qquad (8.33)$$

Numerical example

One calculates in the first place the stiffnesses

$$k_t = 16{,}400\ \mathrm{Nm} \qquad k_c = 1.33 \cdot 10^6\ \mathrm{N/m}$$

It then gives, according to (8.32)

$$\Omega_1^2 = 4.000\ (\mathrm{rad/s})^2 \quad \Omega_2^2 = 13.900\ (\mathrm{rad/s})^2 \quad \Omega_{12}^4 = (6.890)^2 (\mathrm{rad/s})^4$$

One can also plot the circle of angular frequencies in figure 8.9.

Fig. 8.9 Circle of the angular frequencies for example 8.4.1

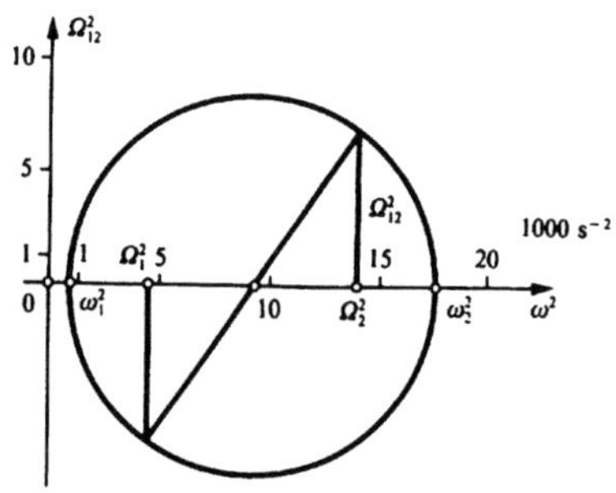


The solutions of equation (8.33) have the values

$$\omega_1^2 = 830\ (\mathrm{rad/s})^2 \qquad \omega_2^2 = 17.600\ (\mathrm{rad/s})^2$$

To summarize, one obtains the following results :

- natural angular frequencies and natural frequencies: $\omega_1 = 28.8$ rad/s, $f_1 = 4.6$ Hz; $\omega_2 = 132.5$ rad/s, $f_2 = 21.1$ Hz
- angular frequencies and frequencies for zero coupling: $\Omega_1 = 66.7$ rad/s, $F_1 = 10.6$ Hz; $\Omega_2 = 118.1$ rad/s, $F_2 = 18.8$ Hz

Comments

In the particular case chosen, the shaft is long (3 m), whereas the cable is relatively short (15 m). Thus, the frequencies $f_1$ and $f_2$ are close enough to each other. On the other hand, if the shaft was short (for example 0,5 m) and if the cable was long (for example 100 m), the frequency $f_1$ would be much lower than $f_2$ . Then the system would no longer present any significant difference from an elementary oscillator consisting simply of the cable and the load.

### 8.4.2 Beats in the free state

A system with two degrees of freedom can present, in the free state, a phenomenon of beats analogous to that of an elementary oscillator being acted upon by two harmonic forces, as examined in paragraph 5.3.1.

Let us return to the relations (8.13) :

$$\begin{cases} x_1 = X_1 \cos(\omega_1 t-\varphi_1) + X_2 \cos(\omega_2 t-\varphi_2) \\ x_2 = \beta_{21} X_1 \cos(\omega_1 t-\varphi_1) + \beta_{22} X_2 \cos(\omega_2 t-\varphi_2) \end{cases} \tag{8.34}$$

One adopts the notation

$$\begin{cases} \omega = \frac{1}{2} (\omega_1 + \omega_2) \qquad \varphi = \frac{1}{2} (\varphi_1 + \varphi_2) \\ \alpha = \frac{1}{2} (\omega_1 - \omega_2) \qquad \psi = \frac{1}{2} (\varphi_1 - \varphi_2) \end{cases} \tag{8.35}$$

By proceeding as before, one finds easily

$$\begin{cases} x_1 = (X_1 + X_2) \cos(\omega t-\varphi) \cdot \cos(\alpha t-\psi) \\ \qquad - (X_1 - X_2) \sin(\omega t-\varphi) \cdot \sin(\alpha t-\psi) \\ x_2 = (\beta_{21} X_1 + \beta_{22} X_2) \cos(\omega t-\varphi) \cdot \cos(\alpha t-\psi) \\ \qquad - (\beta_{21} X_1 - \beta_{22} X_2) \sin(\omega t-\varphi) \cdot \sin(\alpha t-\psi) \end{cases} \tag{8.36}$$

The displacement $x_1$ oscillates at the angular frequency $\omega$ in an envelope which is itself oscillating, at the angular frequency $\alpha$ , between the extreme amplitudes $(X_1 + X_2)$ and $(X_1 - X_2)$ . The displacement $x_2$ does likewise between the amplitudes $(\beta_{21} X_1 + \beta_{22} X_2)$ and $(\beta_{21} X_1 - \beta_{22} X_2)$. This phenomenon of beats, which one can put in the form

$$\begin{cases} x_1 = G_1(\alpha t) \cdot \cos[\omega t - \varphi - \gamma_1(\alpha t)] \\ x_2 = G_2(\alpha t) \cdot \cos[\omega t - \varphi - \gamma_2(\alpha t)] \end{cases} \tag{8.37}$$

is shown for one specific case in figure 8.10.

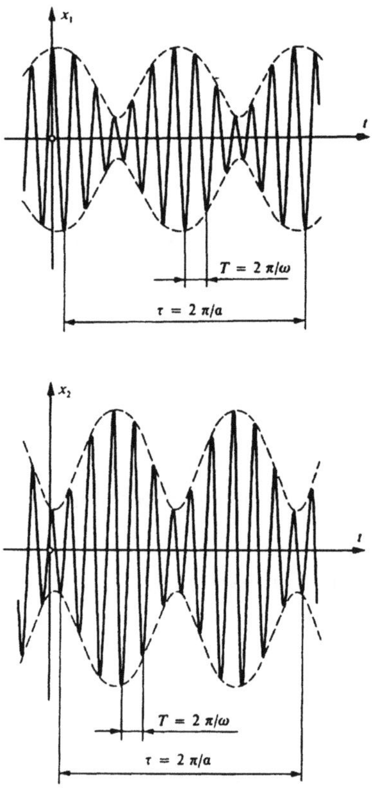


Fig. 8.10 Beats in the free state of a system with two degrees of freedom
$\omega_1 = 60$ rad/s   $X_1 = 0.6$   $\beta_{21} = 1.6$   $\varphi_1 = 0.5$
$\omega_2 = 50$ rad/s   $X_2 = 0.4$   $\beta_{22} = -1.4$   $\varphi_2 = 0.2$

Comments

- One part of the energy of the system is exchanged alternatively between the variables $x_1$ and $x_2$, whilst each mode conserves its own energy.

- For a symmetrical system, for example those of figures 8.2 to 8.4, the coefficients $\beta_1$ and $\beta_2$ take the vales 1 and -1 respectively. If one chose the initial conditions such that $X_1 = X_2$, the displacements $x_1$ and $x_2$ would pass through zero values, at zero speed, for each half period $\tau/2 = \pi/\alpha$. (This is analogous to the situation shown in figure 5.3 case (b), page 90.)

# CHAPTER 9 THE FRAHM DAMPER

## 9.1 Definition and differential equations of the system

A **Frahm damper** is a device which attenuates the vibrations of a mechanical system over a specified range of frequencies. It consists of an oscillating system, known as the auxiliary system, whether dissipative or not, which is attached to the main system, augmenting in this way the number of degrees of freedom and therefore the number of resonances of the complete system. The attenuation of the vibrations of the main system is achieved by the transfer of these to the auxiliary system at the desired frequencies.

One meets it in practice in very varied forms. Figure 9.1 shows a diagram of the principle in its simplest form; the main oscillator $m_1$ , $k_1$ is lacking a natural damper and the secondary oscillator consists of only a single mass $m_2$ . So the complete system has two degrees of freedom. We are going to study, following the method of reference [3], the steady state caused by an harmonic force $F \cos \omega t$ acting on the main mass.

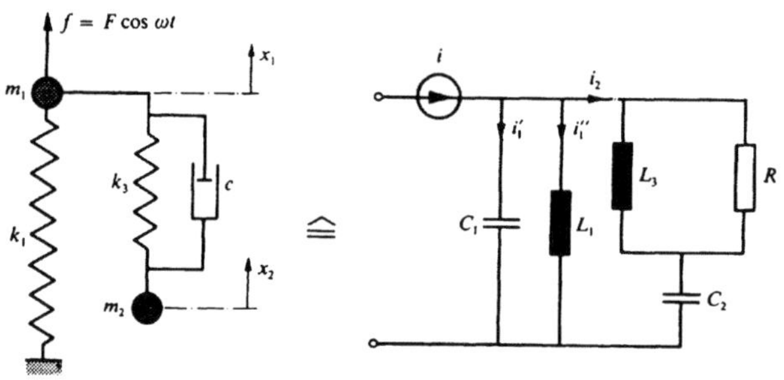


Fig. 9.1 Diagram of the principle of Frahm's damper with the corresponding force-current analogue.

Newton's equations for the system are written

$$\begin{cases} m_1 \ddot{x}_1 + k_1 x_1 + k_3(x_1 - x_2) + c(\dot{x}_1 - \dot{x}_2) = F \cos \omega t \\ m_2 \ddot{x}_2 \qquad\qquad + k_3(x_2 - x_1) + c(\dot{x}_2 - \dot{x}_1) = 0 \end{cases} \tag{9.1}$$

## 9.2 Harmonic steady state

As we will show later for the case of the generalized oscillator, the displacements $x_1$ and $x_2$ in the steady state are harmonic functions with the same angular frequency $\omega$ as the exciting force. We can then write

$$\begin{cases} x_1 = X_1 \cos(\omega t-\varphi_1) \\ x_2 = X_2 \cos(\omega t-\varphi_2) \end{cases}$$

The simplest is then to seek the complex displacements $\underline{X}_1$ and $\underline{X}_2$ of which $x_1$ and $x_2$ are the real parts.

One obtains

$$\begin{cases} \underline{X}_1 = X_1 e^{j(\omega t-\varphi_1)} = X_1 e^{-j\varphi_1} \cdot e^{j\omega t} = \underline{A}_1 e^{j\omega t} \\ \underline{X}_2 = X_2 e^{j(\omega t-\varphi_2)} = X_2 e^{-j\varphi_2} \cdot e^{j\omega t} = \underline{A}_2 e^{j\omega t} \end{cases} \tag{9.2}$$

Likewise, the external force is the real part of the complex force $\underline{F} = F e^{j\omega t}$ . Equations (9.1) become thus

$$\begin{cases} - \omega^2 m_1 \underline{A}_1 + k_1 \underline{A}_1 + k_3(\underline{A}_1 - \underline{A}_2) + j \omega c(\underline{A}_1 - \underline{A}_2) = F \\ - \omega^2 m_2 \underline{A}_2 \qquad\qquad + k_3(\underline{A}_2 - \underline{A}_1) + j \omega c(\underline{A}_2 - \underline{A}_1) = 0 \end{cases}$$

or

$$\begin{cases} \underline{A}_1(- \omega^2 m_1 + k_1 + k_3 + j \omega c) \qquad - \underline{A}_2(k_3 + j \omega c) = F \\ - \underline{A}_1(k_3 + j \omega c) \qquad + \underline{A}_2(- \omega^2 m_2 + k_3 + j \omega c) = 0 \end{cases} \tag{9.3}$$

These equations enable one to determine the complex numbers $\underline{A}_1$ and $\underline{A}_2$ and, as a result, the amplitudes $X_1$ and $X_2$ as well as the phase shifts $\varphi_1$ and $\varphi_2$ . We are only interested in the amplitude of motion of the principal mass. Let us therefore calculate $\underline{A}_1$

$$\underline{A}_1 = F \frac{(k_3-\omega^2 m_2) + j\,\omega\,c}{((k_1-\omega^2 m_1)\,(k_3-\omega^2 m_2) - \omega^2\,m_2\,k_3) + j\,\omega\,c(k_1-\omega^2 m_1 \cdot \omega^2 m_2)} \tag{9.4}$$

It is a complex number of the form

$$\underline{A}_1 = F\,\frac{a + j\,b}{c + j\,d} = F\,\frac{\sqrt{a^2 + b^2}\,e^{j\alpha}}{\sqrt{c^2 + d^2}\,e^{j\beta}}$$

As the first equation (9.2) shows, the modulus of $\underline{A}_1$ is equal to the amplitude $X_1$ . Therefore one has

$$\left(\frac{X_1}{F}\right)^2 = \frac{a^2 + b^2}{c^2 + d^2}$$

which gives, using (9.4)

$$\left(\frac{X_1}{F}\right)^2 = \frac{(k_3-\omega^2 m_2)^2 + \omega^2\,c^2}{((k_1-\omega^2 m_1)\,(k_3-\omega^2 m_2) - \omega^2\,m_2\,k_3)^2 + \omega^2\,c^2(k_1-\omega^2 m_1-\omega^2 m_2)^2} \tag{9.5}$$

We are going to undertake the study of this function in successive steps. The mass and the stiffness of the main oscillator being fixed, we then determine the role of the exciting angular frequency $\omega$ and of the adjustable parameters $m_2$ , $k_3$ , $c$ . We adopt the following notation

$\varepsilon = \frac{m_2}{m_1}$ ratio between the mass of the damper and the main mass,

$\omega_1 = \sqrt{\frac{k_1}{m_1}}$ natural angular frequency of the isolated main oscillator,

$\omega_2 = \sqrt{\frac{k_3}{m_2}}$ natural angular frequency of the isolated undamped secondary oscillator,

$\alpha = \frac{\omega_2}{\omega_1}$ ratio between these natural angular frequencies,

$\beta = \frac{\omega}{\omega_1}$ ratio between the frequency of the driving force and the natural frequency of the main oscillator,

$\eta = \frac{c}{2\, m_2\, \omega_1}$ "hybrid" damping factor

$X_{1s} = \frac{F}{k_1}$ static displacememt of the main mass,

$\mu = \frac{X_1}{X_{1s}}$ dynamic amplification factor of the motion of the main mass.

Using this notation, and after the necessary simplification, relation (9.5) becomes

$$\mu^2 = \frac{4\,\eta^2\,\beta^2 + (\beta^2 - \alpha^2)^2}{4\,\eta^2\,\beta^2(\beta^2(1+\varepsilon) - 1))^2 + (\varepsilon\,\alpha^2\,\beta^2 - (\beta^2 - 1)\,(\beta^2 - \alpha^2))^2} \qquad (9.6)$$

Let us first examine how the amplification factor $\mu$ behaves as a function of the relative angular frequency $\beta$ and with the following values for the parameters :

- $\alpha = 1$ the natural angular frequency of the damper is equal to that of the main oscillator (we shall see below that the optimal value of $\alpha$ is slightly less).
- $\varepsilon = 0.05$ The mass of the oscillator is one twentieth of that of the main system.
- $\eta = 0 \; ; \; 0.1 \; ; \; 0.3 \; ; \; \infty$

One obtains in this way the curves shown in figure 9.2, all of which pass through the two points P and Q , independent of the damping factor. The existence of these points will be demonstrated below.

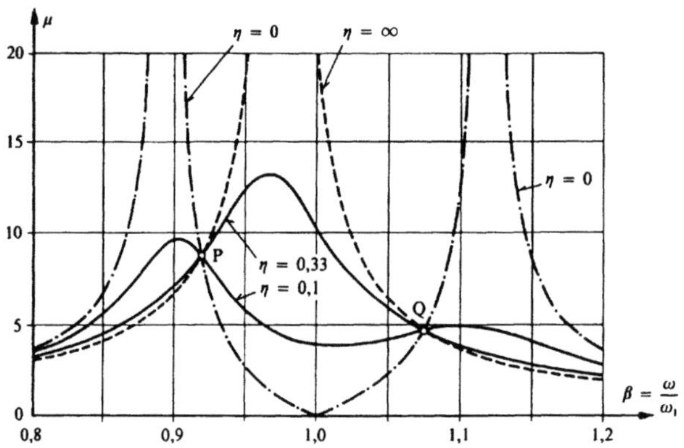


Fig. 9.2 Frahm damper. Dynamic amplification factor as a function of the relative angular frequency

## 9.3 Limiting cases of the damping

Let us examine in more detail the two limiting cases $\alpha = 1$ and $\varepsilon = 0{,}05$ , whilst keeping the values $\eta = 0$ and $\eta = \infty$ .

When the **secondary oscillator is undamped,** the amplification factor takes the simple, absolute, form,

$$\mu = \left| \frac{\beta^2 - 1}{0.05\ \beta^2 - (\beta^2 - 1)^2} \right| \tag{9.7}$$

It becomes infinite for the values of $\beta$ which make the denominator zero ($\beta'$ and $\beta''$ in figure 9.3). On the other hand, for $\beta = 1$, $\mu = 0$ : the main mass remains at rest while all the force is absorbed by the secondary oscillator. This particular circumstance appears clearly in the electrical circuit diagram; if the resistance R becomes infinite ($c = 0 \Rightarrow R = \infty$) and the secondary circuit $L_3$ , $C_2$ is tuned, the main circuit is short-circuited, which implies $i_1' = i_1'' = 0 \Rightarrow x_1 = 0$ .

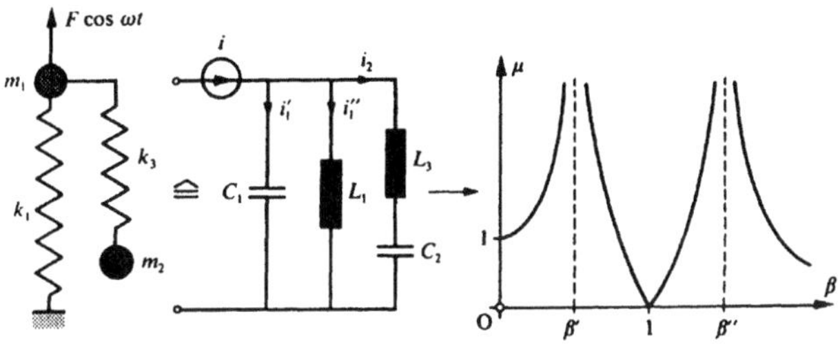


Fig. 9.3 Frahm damper and amplification factor in the case $\eta = 0$, $\alpha = 1$ and $\varepsilon = 0.05$

Contrary to what one could think at first sight, the case $\eta = 0$ is not generally favorable in practice, even for a machine whose steady state corresponds to $\beta = 1 \Rightarrow \omega = \omega_1 = \omega_2$ . Indeed, when accelerating or decelerating the machine, one inevitably passes through some very large amplitudes in the neighbourhood of the angular frequency $\omega' = \beta'\omega_1$ . Moreover, if $\eta = 0$ and $\beta = 1$ , the amplitudes of the secondary mass, which we have not calculated here, can become excessive.

If the **resistance c is infinite,** the masses $m_1$ and $m_2$ are linked rigidly and the system degenerates into an undamped elementary oscillator whose mass is equal to $m_1(1 + \varepsilon)$ . In the electrical diagram (figure 9.4), the capacities $C_1$ and $C_2$ are in parallel for $c = \infty \Rightarrow R = 0$ .

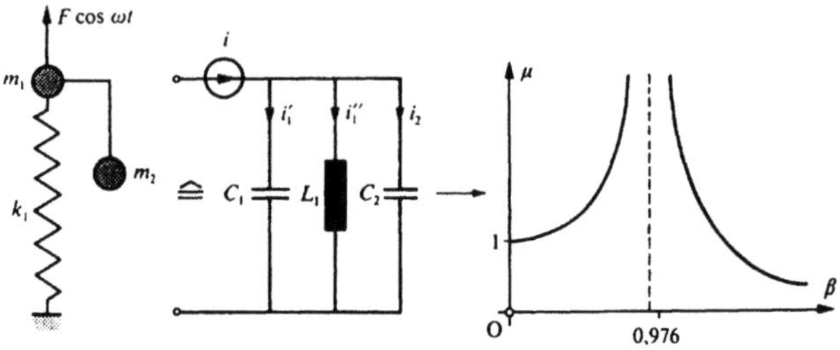


Fig. 9.4 Frahm damper and amplification factor in the case $\eta = \infty$, $\alpha = 1$ and $\varepsilon = 0.05$

The amplification factor becomes, according to (9.6) :

$$\mu = \frac{1}{|\beta^2(1 + 0.05) - 1|} \tag{9.8}$$

It is infinite for $\beta = 1/\sqrt{1.05} = 0.976$ .

In summary, the coefficient $\mu$ can become infinite in the two limiting cases $\eta = 0$ and $\eta = \infty$ . In general, the principal object of the damping system is to limit the movement of the main mass $m_1$ for as large a range of frequencies as possible. We shall assume this to be the case for the rest of this study and shall say, when this condition is fulfilled, that the **damping is optimal**. Of course, the optimization could be determined by other criteria.

## 9.4 Optimization of the Frahm damper

Let us return to figure 9.2. The parameters $\alpha$ and $\varepsilon$ having constant values (1 and 0.05 respectively), the points P and Q have a fixed position, as we have already indicated. That being the case, the most favourable value of $\eta$ which one could choose in such a case is that which corresponds to a curve $\mu(\beta)$ having a horizontal tangent at the highest point, that is to say point P . However, one can do better by lowering the point P .

In fact, one easily verifies that if the ratio $\alpha$ of the natural angular frequencies varies, the points P and Q are displaced along the curve $\eta = 0$ . Since one rises if the other descends, the optimal situation is reached when they are both at the same height. One then chooses a curve $\mu(\eta)$ passing through one with a horizontal tangent.

Let us now show, from relation (9.6), that there effectively exists two values of $\beta$ for which $\mu$ is independant of $\eta$ . One can write

$$\mu^2 = \frac{A\ \eta^2 + B}{C\ \eta^2 + D}$$

In order that $\mu$ is independant of $\eta$ , it is necessary that $A/C = B/D$ which leads to

$$\frac{1}{(\beta^2(1 + \varepsilon) - 1)^2} = \frac{(\beta^2 - \alpha^2)^2}{(\varepsilon\ \alpha^2\ \beta^2 - (\beta^2 - 1)\ (\beta^2 - \alpha^2))^2}$$

which gives

$$\varepsilon\ \alpha^2\ \beta^2 - (\beta^2 - 1)\ (\beta^2 - \alpha^2) = \pm\ (\beta^2 - \alpha^2)\ (\beta^2(1 + \varepsilon) - 1)$$

With the minus sign, this condition becomes

$$\varepsilon\ \beta^2 \cdot \beta^2 = 0 \quad \Rightarrow \quad \beta = 0 \quad \Rightarrow \quad \omega = 0$$

This result shows that all the curves tend towards the point $\mu = 1$ for $\beta \to 0$ . It does not concern the points P and Q sought.

With the plus sign, the condition can be written

$$\beta^4(2 + \varepsilon) - \beta^2(2 + 2\ \alpha^2(1 + \varepsilon)) + 2\ \alpha^2 = 0$$

or again

$$\beta^4 - 2\,\beta^2\,\frac{1 + \alpha^2(1 + \varepsilon)}{2 + \varepsilon} + \frac{2\,\alpha^2}{2 + \varepsilon} = 0 \tag{9.9}$$

This is a bi-quadratic equation whose two positive roots $\beta_1$ and $\beta_2$, both functions of $\alpha$ and $\varepsilon$, correspond to the points P and Q.

We should like the amplification factor to have same value at the points P and Q, $\mu(P) = \mu(Q)$. It is useless to introduce $\beta_1^2$ then $\beta_2^2$ into equation (9.6) for we know that at points P and Q the value of $\mu$ is independent of the damping factor $\eta$. One can therefore choose $\eta$ in such a way that the equation is simplified to the maximum. This is the case for $\eta = \infty$ and (9.6) becomes

$$\mu = \frac{1}{|1 - \beta^2(1 + \varepsilon)|} \tag{9.10}$$

Since P is on the side $\mu > 0$ and Q of the side $\mu < 0$, it is necessary to write

$$\mu(\beta_1^2) = -\,\mu(\beta_2^2)$$

$$\frac{1}{1 - \beta_1^2(1 + \varepsilon)} = \frac{-1}{1 - \beta_2^2(1 + \varepsilon)}$$

$$1 - \beta_1^2(1 + \varepsilon) = -1 + \beta_2^2(1 + \varepsilon)$$

$$\beta_1^2 + \beta_2^2 = \frac{2}{1 + \varepsilon} \tag{9.11}$$

Let us calculate the sum of the roots of equation (9.9)

$$\beta_1^2 + \beta_2^2 = 2\,\frac{1 + \alpha^2(1 + \varepsilon)}{2 + \varepsilon} \tag{9.12}$$

Whence, by equating (9.11) and (9.12)

$$\frac{2}{1 + \varepsilon} = 2\,\frac{1 + \alpha^2(1 + \varepsilon)}{2 + \varepsilon}$$

$$\alpha^2 = \frac{1}{(1 + \varepsilon)^2}$$

In the extraction of the root, only the positive sign must be considered, since the ratio $\alpha = \omega_2/\omega_1$ is essentially positive. Thus

$$\alpha = \frac{1}{1 + \varepsilon} \tag{9.13}$$

Let us recall that for this value of $\alpha$ the points P and Q are at the same height. Thus one sees that the optimal solution is obtained when the natural angular frequency of the damper is slightly less than that of the main system (ie $\varepsilon$ is small).

It is necessary to determine again the value of $\mu$ which one obtains when the condition (9.13) is fulfilled. For that, let us consider the two curves having a horizontal tangent at the points P and Q respectively (figure 9.5). One can assume them to be horizontal between these points without a large error. Consequently, one transfers the roots of equation (9.9) into relation (9.10).

Let us first calculate the roots of (9.9) when the condition (9.13) is satisfied.

$$\beta^4 - 2\,\beta^2 \frac{1}{1 + \varepsilon} + \frac{2}{(2 + \varepsilon)\,(1 + \varepsilon)^2} = 0$$

$$\beta^2 = \frac{1}{1 + \varepsilon}\left(1 \pm \sqrt{\frac{\varepsilon}{2 + \varepsilon}}\right)$$

Let us introduce these values into (9.10)

$$\mu = \frac{1}{1 - \left(1 \pm \sqrt{\dfrac{\varepsilon}{2 + \varepsilon}}\right)} = \frac{1}{\pm \sqrt{\dfrac{\varepsilon}{2 + \varepsilon}}}$$

$$\mu = \sqrt{\frac{2 + \varepsilon}{\varepsilon}} \tag{9.14}$$

## 9.5 Examples of applications

Let us first consider an optimal damper for which $\varepsilon = 1/4$ . The optimal value of $\alpha$ is, from (9.13) :

$$\alpha = \frac{1}{1 + 1/4} = \frac{4}{5} = 0,8$$

The abscissae of the points P and Q have the values

$$\beta^2 = \frac{1}{1 + 0.25} \left(1 \pm \sqrt{\frac{0.25}{2 + 0.25}}\right) \text{, which gives}$$

$$\beta_1 = 0.730 \qquad \beta_2 = 1.033$$

Figure 9.5 shows the limiting curves $\eta = 0$ and $\eta = \infty$ in this particular case, as well as the two curves having horizontal tangents at P and Q respectively.

From (9.14), the amplification factor has the value

$$\mu = \sqrt{\frac{2 + 0.25}{0.25}} = 3$$

Let us compare this result with the case of a damper for which $\omega_2 = \omega_1$ , so $\alpha = 1$ .

Equation (9.9) becomes then

$$\beta^4 - 2\,\beta^2 + \frac{2}{2 + \varepsilon} = 0$$

whence

$$\beta^2 = 1 \pm \sqrt{1 - \frac{2}{2 + \varepsilon}} = 1 \pm \sqrt{\frac{\varepsilon}{2 + \varepsilon}}$$

In all practical cases, $\mu$ is largest for the smallest value of $\beta$. Let us take therefore

$$\beta^2 = 1 - \sqrt{\frac{\varepsilon}{2 + \varepsilon}} \qquad (9.15)$$

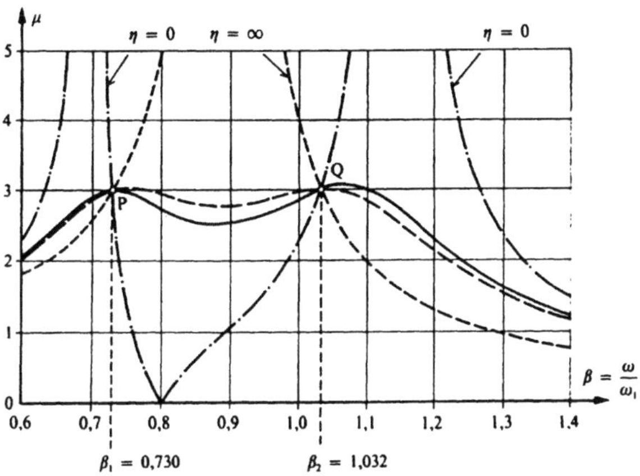


Fig. 9.5 Optimal Frahm damper. Dynamic amplification factor as a function of the relative angular frequency

and let us introduce this value into (9.10) to calculate $\mu$

$$\mu = \frac{1}{\left| - \varepsilon + (1 + \varepsilon) \sqrt{\frac{\varepsilon}{2 + \varepsilon}} \right|} \tag{9.16}$$

In the particular case considered ($\varepsilon = 1/4$), $\mu$ becomes

$$\mu = \frac{1}{- 0.25 + 1.25 \sqrt{\frac{0.25}{2.25}}} = 6$$

that is to say double the value obtained in the optimal case.

We are going to calculate the damping factor corresponding to the curve of figure 9.5 having a horizontal tangent at the point P.

Let us return to relation (9.6) giving the square of the dynamic amplification factor

$$\mu^2 = \frac{N}{D} = \frac{4\,\eta^2\,\beta^2 + (\beta^2 - \alpha^2)^2}{4\,\eta^2\,\beta^2\,(\beta^2\,(1 + \varepsilon) - 1)^2 + (\varepsilon\,\alpha^2\,\beta^2 - (\beta^2 - 1)\,(\beta^2 - \alpha^2))^2}$$

It is a question of determining $\eta$ in such a way that the partial derivative $\partial\mu/\partial\beta$ is zero for $\beta = \beta_1$, which is equivalent to the following condition, which is much simpler to express

$$\left(\frac{\partial\mu^2}{\partial\beta^2}\right)_{\beta_1^2} = 0$$

One has in the first place

$$\frac{\partial\mu^2}{\partial\beta^2} = \frac{1}{D^2}\left(D\,\frac{\partial N}{\partial\beta^2} - N\,\frac{\partial D}{\partial\beta^2}\right) = \frac{1}{D}\left(\frac{\partial N}{\partial\beta^2} - \frac{N}{D}\,\frac{\partial D}{\partial\beta^2}\right) = 0$$

The condition to satisfy is in the same way, with $N/D = \mu^2(P) = \lambda$

$$\frac{\partial N}{\partial\beta^2} - \lambda\,\frac{\partial D}{\partial\beta^2} = 0 \qquad (9.17)$$

Differentiating gives

$$\frac{\partial N}{\partial\beta^2} = 4\,\eta^2 + 2(\beta^2 - \alpha^2) \qquad (9.18)$$

$$\begin{aligned}\frac{\partial D}{\partial\beta^2} &= 4\,\eta^2(\beta^2(1 + \varepsilon) - 1)^2 + 8\,\eta^2\,\beta^2(\beta^2(1 + \varepsilon) - 1)\,(1 + \varepsilon) \\ &\quad + 2(\varepsilon\,\alpha^2\,\beta^2 - (\beta^2 - 1)\,(\beta^2 - \alpha^2))\,(\varepsilon\,\alpha^2 - (\beta^2 - \alpha^2) - (\beta^2 - 1))\end{aligned} \qquad (9.19)$$

With $\lambda = 9$, $\varepsilon = 0.25$, $\beta^2 = 0.533$ and taking account of (9.18) and (9.19), equation (9.17) is reduced to

$$\eta^2 - 0.04267 = 0 \quad \Rightarrow \quad \eta = 0.207$$

Thus, the curve $\mu(\beta)$ has a horizontal tangent at the point P if $\eta = 0.207$ . By proceeding in the same way, one shows that $\mu(\beta)$ has a horizontal tangent at the point Q when $\eta = 0.231$ . That means that in practice one can choose $0.21 \leqslant \eta \leqslant 0.23$ .

## 9.6 The Lanchester damper

Finally, let us consider the case of Lanchester damper in which the secondary mass and the main mass are only linked by a dash-pot, without spring (figure 9.6).

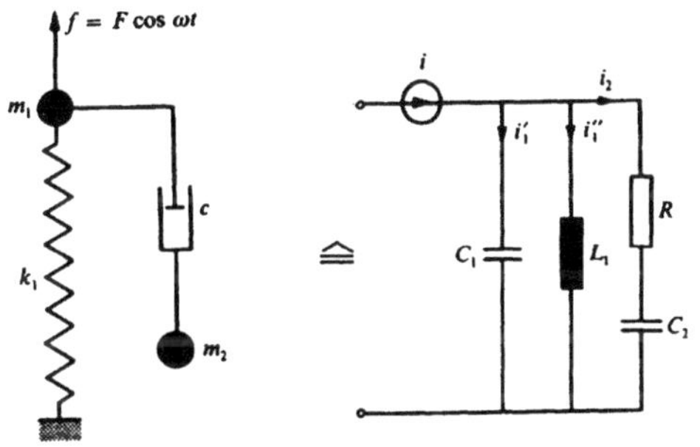


Fig. 9.6 The Lanchester damper

That comes back to saying that $k_3 = 0$ , hence $\omega_2 = 0$ , $\alpha = 0$ and equation (9.9) becomes

$$\beta^4 - 2\,\beta^2 \frac{1}{2 + \varepsilon} = 0$$

The first solution $\beta_1^2 = 0$ means that the point P coincides with the point common to all the curves, which has the coordinates $\beta = 0$ and $\mu = 1$ (figure 9.7). The second solution has the value

$$\beta_2^2 = \frac{2}{2 + \varepsilon} \quad \Rightarrow \quad \beta_2 = \sqrt{\frac{2}{2 + \varepsilon}} \tag{9.20}$$

It gives the abscissa of the point Q for which the amplification factor becomes, using (9.10),

$$\mu = \frac{1}{\left|1 - \frac{2}{2 + \varepsilon}(1 + \varepsilon)\right|} = 1 + \frac{2}{\varepsilon} \tag{9.21}$$

Always in the specific case $\varepsilon = 1/4$, this gives $\mu = 9$, that is to say a value three times that of optimized damper.

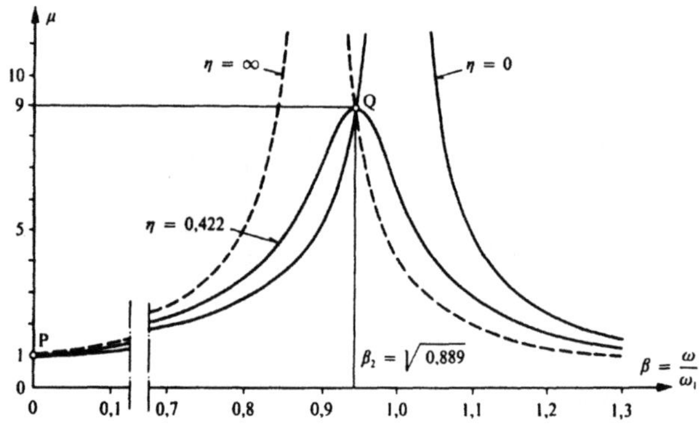


Fig. 9.7 The Lanchester damper - Dynamic amplification factor as a function of the relative angular frequency

In general, the Lanchester damper is not realized with a viscous resistance but with a dry frictional (Coulomb) resistance. The problem is then considerably complicated and one can show that the maximum amplification coefficient has the approximate value

$$\mu = \frac{\pi^2}{4\,\varepsilon} \tag{9.22}$$

which gives, in the particular case considered

$$\mu = \frac{\pi^2}{4 \times 0.25} = 9.87 \approx 10$$

This value is a bit less favourable than that found for the Lanchester damper with viscous friction (10 in place of 9).

# CHAPTER 10 THE CONCEPT OF THE GENERALIZED OSCILLATOR

## 10.1 Definition and energetic forms of the generalized oscillator

A linear mechanical system consisting of any finite number of degrees of freedom, $n$, and whose behaviour is governed by the following second order differential equation

$$[M]\,\ddot{\vec{x}} + [C]\,\dot{\vec{x}} + [K]\,\vec{x} = \vec{f}(t) \qquad (10.1)$$

is called a discrete **general** linear **oscillating system** - or more simply a generalized oscillator.

In this equation, as in section 8.1, $\vec{x}$, $\dot{\vec{x}}$ and $\ddot{\vec{x}}$ are the vectors, of displacements, velocities and accelerations respectively. As for the vector $\vec{f}(t)$, it represents the external forces acting on the system.

The components $x_i$ of the vector $\vec{x}$ are called generalized coordinates, because their physical dimensions are in general different (lengths, angles, volumes, etc.). The same comment applies to the derivatives $\dot{x}_i$ and $\ddot{x}_i$ as well as to the forces $f_i(t)$.

The matrices [M], [C] and [K], which one assumes to be symmetrical, are called the **mass matrix, damping matrix** and **stiffness matrix** respectively.

The mechanical problems leading to an equation of the type (10.1) are very varied. They concern most often the study of small movements of systems belonging to one of the two following categories.

- systems of rigid solids (those which are assumed to keep their shape), being acted upon by elastic forces and linear resistive forces;
- deformable continous systems (ie beams, plates, shells, any structure) made discrete, that is to say replaced in an approxi-

mate way, by a system comprising only a limited number of degrees of freedom, using numerical or experimental methods.

The concept of the generalized oscillator, as one comes to define it, implies that the corresponding energetic forms are given by the expressions below.

The **kinetic energy** depends only on the generalized velocities and has the value

$$T = \frac{1}{2} \dot{\vec{x}}^T [M] \dot{\vec{x}} \tag{10.2}$$

This is a positive definite symmetrical quadratic form for the generalized velocities; it is only zero for $\dot{\vec{x}} = \vec{0}$ .

Therefore, the matrix [M] is equally positive definite symmetrical. That means that it satisfies Silvester's criterion; the determinant of [M] and the determinants of all the diagonal minors must be positive.

$$\Delta_n = |M| = \begin{vmatrix} m_{11} & \cdots & m_{1n} \\ \cdot & & \cdot \\ \cdot & & \cdot \\ \cdot & & \cdot \\ m_{m1} & \cdots & m_{nn} \end{vmatrix} > 0 \quad , \quad \Delta_{n-1} = \begin{vmatrix} m_{11} & \cdots & m_{1,n-1} \\ \cdot & & \cdot \\ \cdot & & \cdot \\ \cdot & & \cdot \\ m_{n-1,1} & \cdots & m_{n-1,n-1} \end{vmatrix} > 0$$

$$, \ldots , \Delta_2 = \begin{vmatrix} m_{11} & m_{12} \\ & \\ m_{21} & m_{22} \end{vmatrix} > 0 \quad , \quad \Delta_1 = m_{11} > 0$$

The **potential energy,** of elastic origin, is given by the expression

$$V = \frac{1}{2} \vec{x}^T [K] \vec{x} \tag{10.3}$$

It is a positive semi-definite symmetrical quadratic form of the generalized displacements. It is said to be semi-definite because it can be zero for a value of $\vec{x}$ different from zero.

The total half-power consumed by the system is called **Rayleigh's dissipation function.**

$$W = \frac{1}{2} \dot{\vec{x}}^T [C] \dot{\vec{x}} \tag{10.4}$$

This function, introduced by Lord Rayleigh, is a positive semi-definite symmetrical quadratic form of the generalized velocities.

We are now going to show that by differentiating relations (10.2) to (10.4) one finds again in fact the differential equation (10.1).

## 10.2 Differentiation of a symmetrical quadratic form · Equations of Lagrange

Before pursuing the study of the oscillator, it is necessary to carry out the following preparatory calculation. One considers a symmetrical quadratic form of the variables $x_i$

$$Q = \vec{x}^T [S] \vec{x} = \sum_i^n \sum_j^n s_{ij} x_i x_j \qquad s_{ij} = s_{ji} \tag{10.5}$$

One calculates the partial derivatives $\frac{\partial Q}{\partial x_1}, \ldots, \frac{\partial Q}{\partial x_n}$ and one expresses the result in matrix form.

If $x_k$ is one of the $x_i$ , the quadratic form can be written

$$Q = \underbrace{\sum_{i \neq k}^n s_{ik} x_i x_k}_{(n-1) \text{ terms}} + \underbrace{\sum_{j \neq k}^n s_{kj} x_k x_j}_{(n-1) \text{ terms}} + \underbrace{s_{kk} x_k^2}_{1 \text{ term}} + \underbrace{Q'}_{(n-1)^2 \text{ terms}} \tag{10.6}$$

In this expression, the term $Q'$ does not depend on the index $k$. Differentiation gives as follows

$$\frac{\partial Q}{\partial x_k} = \sum_{i \neq k}^n s_{ik} x_i + \sum_{j \neq k}^n s_{kj} x_j + 2 s_{kk} x_k = \sum_i^n s_{ik} x_i + \sum_j^n s_{kj} x_j$$

since $s_{ik} = s_{kj}$ if $i = j$, the two sums are equal

$$\frac{\partial Q}{\partial x_k} = 2 \sum_i^n s_{ik} x_i = 2 \sum_j^n s_{kj} x_j$$

By making the index k vary between 1 and n, one obtains

$$\left.\begin{matrix} \dfrac{\partial Q}{\partial x_1} = 2 \sum_j^n s_{1j} x_j \\ \vdots \\ \dfrac{\partial Q}{\partial x_n} = 2 \sum_j^n s_{nj} x_j \end{matrix}\right. \Rightarrow \begin{Bmatrix} \dfrac{\partial Q}{\partial x_1} \\ \vdots \\ \dfrac{\partial Q}{\partial x_n} \end{Bmatrix} = 2\,[S]\,\vec{x}$$

One can define the matrix operator

$$\left\{\frac{\partial}{\partial x_i}\right\} = \begin{Bmatrix} \dfrac{\partial}{\partial x_1} \\ \vdots \\ \dfrac{\partial}{x_n} \end{Bmatrix} \qquad (10.7)$$

To summarize, if Q is a symmetrical quadratic form

$$\left\{\frac{\partial Q}{\partial x_i}\right\} = 2\,[S]\,\vec{x} \qquad (10.8)$$

If, as above, one designates the kinetic energy by T, the potential energy by V and the dissipation function by W, the equations of Lagrange for a dissipative system with n degrees of freedom can be written.

$$\frac{d}{dt}\left(\frac{\partial T}{\partial \dot{x}_k}\right) - \frac{\partial T}{\partial x_k} + \frac{\partial V}{\partial x_k} + \frac{\partial W}{\partial \dot{x}_k} = f_k(t) \qquad k = 1, \ldots, n \qquad (10.9)$$

There are n of them, but the number can be reduced to one equation by using the matrix operator defined above. It gives in this way

$$\frac{d}{dt}\left\{\frac{\partial T}{\partial \dot{x}_i}\right\} - \left\{\frac{\partial T}{\partial x_i}\right\} + \left\{\frac{\partial V}{\partial x_i}\right\} + \left\{\frac{\partial W}{\partial \dot{x}_1}\right\} = \vec{f}(t) \tag{10.10}$$

In the particular case of the generalized oscillator, taking into account the hypotheses made at the beginning of this chapter, the kinetic energy does not depend on the displacements and the above equation simplifies to

$$\frac{d}{dt}\left\{\frac{\partial T}{\partial \dot{x}_i}\right\} + \left\{\frac{\partial V}{\partial x_i}\right\} + \left\{\frac{\partial W}{\partial \dot{x}_i}\right\} = \vec{f}(t) \tag{10.11}$$

Let us recall that T , V and W have the values

$$T = \frac{1}{2} \dot{\vec{x}}^T [M] \dot{\vec{x}}$$

$$V = \frac{1}{2} \vec{x}^T [K] \vec{x}$$

$$W = \frac{1}{2} \dot{\vec{x}}^T [C] \dot{\vec{x}}$$

By using the result (10.8), equation (10.11) becomes

$$\frac{d}{dt} [M] \dot{\vec{x}} + [K] \vec{x} + [C] \dot{\vec{x}} = \vec{f}t)$$

As the coefficients of the matrices are independent of time, one can carry out the differentiation of the first term

$$[M] \ddot{\vec{x}} + [C] \dot{\vec{x}} + [K] \vec{x} = \vec{f}(t)$$

In this way, one finds again the Newtonian equation of the system in matrix form. It is clearly the same as equation (10.1).

## 10.3 Examination of particular cases

### 10.3.1 Energetic forms of the oscillator with two degrees of freedom

The standard diagram of an oscillator with two degrees of freedom, shown in figure 8.1, is reproduced below for convenience.

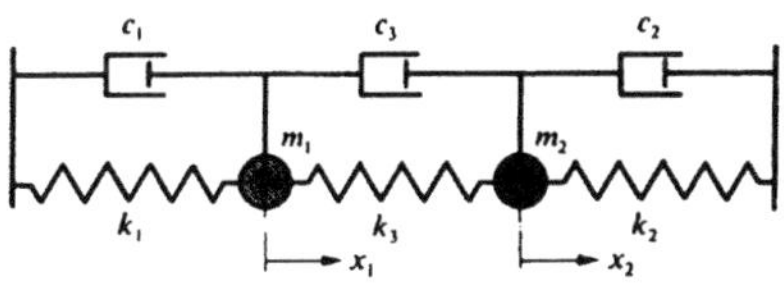


Fig. 10.1 Oscillator with two degrees of freedom

The kinetic energy of the system has the value

$$T = \frac{1}{2} m_1 \dot{x}_1^2 + \frac{1}{2} m_2 \dot{x}_2^2$$

That is to say, in matrix form

$$T = \frac{1}{2} \{\dot{x}_1 \ \dot{x}_2\} \begin{bmatrix} m_1 & 0 \\ 0 & m_2 \end{bmatrix} \begin{Bmatrix} \dot{x}_1 \\ \dot{x}_2 \end{Bmatrix} = \frac{1}{2} \dot{\vec{x}}^T [M] \dot{\vec{x}}$$

One finds again in the same way relation (10.2).

For the potential energy, it gives successively

$$V = \frac{1}{2} (k_1 x_1^2 + k_3(x_1 - x_2)^2 + k_2 x_2^2)$$

$$V = \frac{1}{2} ((k_1 + k_3) x_1^2 - 2 k_3 x_1 x_2 + (k_2 + k_3) x_2^2)$$

$$V = \frac{1}{2} \{x_1 \ x_2\} \begin{bmatrix} k_1 + k_3 & -k_3 \\ -k_3 & k_2 + k_3 \end{bmatrix} \begin{Bmatrix} x_1 \\ x_2 \end{Bmatrix} = \frac{1}{2} \vec{x}^T [K] \vec{x}$$

Which is relation (10.3).

The force $c_1 \dot{x}_1$ , due to the resistance $c_1$ , dissipates power $c_1 \dot{x}_1 \cdot \dot{x}_1 = c_1 \dot{x}_1^2$ . Likewise, the power dissipated in $c_3$ and $c_2$ are $c_3(\dot{x}_1 - \dot{x}_2)^2$ and $c_2 \dot{x}_2^2$ respectively. The half-sum of these powers represents Rayleigh's dissipation function.

$$W = \frac{1}{2} (c_1 \dot{x}_1^2 + c_3(\dot{x}_1 - \dot{x}_2)^2 + c_2 \dot{x}_2^2)$$

$$W = \frac{1}{2} ((c_1 + c_3) \dot{x}_1^2 - 2 c_3 \dot{x}_1 \dot{x}_2 + (c_2 + c_3) \dot{x}_2^2)$$

$$W = \frac{1}{2} \{\dot{x}_1 \ \dot{x}_2\} \begin{bmatrix} c_1 + c_3 & - c_3 \\ - c_3 & c_2 + c_3 \end{bmatrix} \begin{Bmatrix} \dot{x}_1 \\ \dot{x}_2 \end{Bmatrix} = \frac{1}{2} \dot{\vec{x}}^T [C] \dot{\vec{x}}$$

Relation (10.4) is established in this way.

One confirms that the matrices [M], [C] and [K] are symmetrical and that they clearly have the values found from Newton's equations (8.4). On the other hand, one notes that the structure of [C] is the same as that of [K] . This similarity is not general but never-the-less frequent. Conversely, the structure of [M] is intrinsically different.

### 10.3.2 Potential energy of a linear elastic system

One considers a linear elastic system, subject to $n$ generalized forces $Q_1, \ldots, Q_n$ . The point of application $A_i$ of the force $Q_i$ is displaced by $A_i'$ (figure 10.2). Let us designate by $x_i$ the component of $A_iA_i'$ along $Q_i$ , and by $b_i$ the component in the orthogonal plane. The deformed configuration of the system, from the initial configuration ($Q_i'=0$) is defined by the set of displacements $x_i$ , $b_i$ .

By hypothesis, the forces are linear functions of the displacements and one can write

$$Q_i = \sum_j^n k_{ij} x_j \tag{10.12}$$

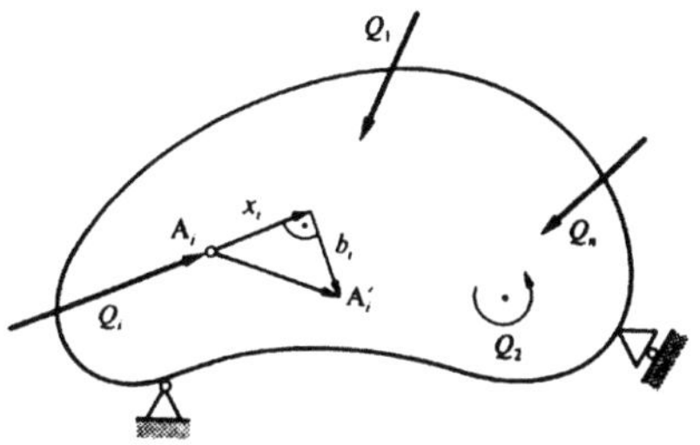


Fig. 10.2 Linear elastic system subject to n forces $Q_i$

One knows that the reciprocal stiffnesses are equal (Maxwell-Betti theorem)

$$k_{ij} = k_{ji} \tag{10.13}$$

The system being linear, the potentiel energy of deformation is equal to the half-sum of the products between the forces and the displacements in their directions (Clapeyron's equation)

$$V = \frac{1}{2} \sum_{i}^{n} Q_i \, x_i \tag{10.14}$$

By replacing the $Q_i$ with their values (10.12), V appears as a symmetric quadratic form of the displacements

$$V = \frac{1}{2} \sum_{i}^{n} \sum_{j}^{n} k_{ij} \, x_i \, x_j \tag{10.15}$$

Let us again write the results (10.12), (10.14) and (10.15) in matrix form

$$\vec{Q} = [K]\,\vec{x} \tag{10.16}$$

$$V = \frac{1}{2}\,\vec{Q}^T\,\vec{x} = \frac{1}{2}\,\vec{x}^T\,\vec{Q} \tag{10.17}$$

$$V = \frac{1}{2}\,\vec{x}^T\,[K]\,\vec{x}$$

To summarize, in the particular case examined, $V$ is clearly of the form (10.3).

The inversion of (10.16) enables us to define the **flexibility matrix** $[\alpha]$, also known as the **matrix of influence coefficients** which is sometimes easier to determine in practice.

$$\vec{x} = [K]^{-1}\,\vec{Q} = [\alpha]\,\vec{Q} \tag{10.18}$$

The matrix $[\alpha]$ is symmetric since $[K]$ is symmetric. Regarding the potential energy, it becomes a symmetric quadratic form of the forces

$$V = \frac{1}{2}\,\vec{Q}^T\,[\alpha]\,\vec{Q} \tag{10.19}$$

### 10.3.3 Kinetic energy of a system of point masses

It is easy to show, as we are going to below, that the kinetic energy of a system of point masses is a positive definite symmetric quadratic form of the generalized velocities. Let us therefore examine a system of $r$ point masses $m_\alpha$ (figure 10.3), linked between themselves and to the planes of an inertial reference system by elastic forces and by viscous damping forces. One makes the hypothesis that the system is acted upon by $\ell$ holonomic constraints, which are bilateral and independant of time. It therefore has $n = 3\,r - \ell$ degrees of freedom.

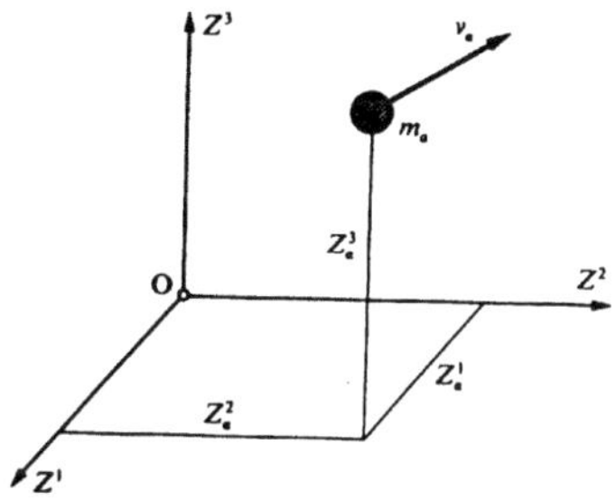


Fig. 10.3 System of r point masses

Let $z_\alpha^k$ (k=1,2,3) be the cartesian coordinates of the mass $m_\alpha$ in the reference frame chosen. The velocity of this mass being $V_\alpha$, one can write

$$v_\alpha^2 = \sum_k^3 (\dot{z}_\alpha^k)^2 \tag{10.20}$$

The total kinetic energy has the value

$$T = \frac{1}{2} \sum_\alpha^r m_\alpha \, v_\alpha^2$$

which gives, by using relation (10.20)

$$T = \frac{1}{2} \sum_\alpha^r \sum_k^3 m_\alpha (\dot{z}_\alpha^k)^2 \tag{10.21}$$

The configuration of the system can be described by n generalized coordinates $x_i (i=1,\ldots,n)$ . Because of the hypothesis adopted on the nature of the constraints, the coordinates $z_\alpha^k$ are not explicite functions of time and only depend on time via the time-dependence of the $x_i$ . One has then

$$z_\alpha^k = z_\alpha^k(x_1, \ldots, x_i, \ldots, x_n)$$

$$dz^k_\alpha = \sum_i^n \frac{\partial z^k_\alpha}{\partial x_i} dx_i \qquad => \qquad \dot{z}^k_\alpha = \sum_i^n \frac{\partial z^k_\alpha}{\partial x_i} \dot{x}_i \tag{10.22}$$

In order to simplify the notation, let us adopt the convention

$$h^k_{\alpha i} = \frac{\partial z^k_\alpha}{\partial x_i} \tag{10.23}$$

The kinetic energy becomes thus

$$T = \frac{1}{2} \sum_\alpha^r \sum_k^3 m_\alpha (\sum_i^n h^k_{\alpha i} \dot{x}_i)^2$$

By using two indices in place of one alone, the square of the last sum can be put in the form

$$(\sum_i^n h^k_{\alpha i} \dot{x}_i)^2 = \sum_i^n \sum_j^n h^k_{\alpha i} h^k_{\alpha j} \dot{x}_i \dot{x}_j$$

with, naturally, $h^k_{\alpha i} = h^k_{\alpha j}$ if $i = j$ .

It gives thus

$$T = \frac{1}{2} \sum_\alpha^r \sum_k^3 m_\alpha \sum_i^n \sum_j^n h^k_{\alpha i} h^k_{\alpha j} \dot{x}_i \dot{x}_j$$

$$T = \frac{1}{2} \sum_i^n \sum_j^n \dot{x}_i \dot{x}_j \sum_\alpha^r \sum_k^3 m_\alpha h^k_{\alpha i} h^k_{\alpha j} \tag{10.24}$$

One calls the quantities defined as follows **generalized masses**, or coefficients of inertia,

$$m_{ij} = \sum_\alpha^r \sum_k^3 m_\alpha h^k_{\alpha i} h^k_{\alpha j} \tag{10.25}$$

Since $h^k_{\alpha i} = h^k_{\alpha j}$ if $i = j$ , the reciprocal generalized masses are equal : $m_{ij} = m_{ji}$ .

The kinetic energy becomes finally

$$T = \frac{1}{2} \sum_i^n \sum_j^n m_{ij} \dot{x}_i \dot{x}_j \tag{10.26}$$

Written in the matrix form, this result clearly gives again relation (10.2)

$$T = \frac{1}{2} \dot{\vec{x}}^T [M] \dot{\vec{x}}$$

# CHAPTER 11 FREE STATE OF THE CONSERVATIVE GENERALIZED OSCILLATOR

## 11.1 Introduction

As we have done for the elementary oscillator, we are going to assume at first that the generalized oscillator is not subject to any external force and that there is no damping ($\vec{f}(t) = \vec{0}$ , $[C] = [0]$) . With these conditions, the solution of the differential system is the conservative free state.

Knowledge of the conservative free state, which is very much easier to establish than for the dissipative free state, is sufficient for many practical problems. The natural frequencies, in particular, are only slighty influenced by damping (except for exceptionally high values of that). Let it be clearly understood, this is different from the study of forced states, for which damping plays an essential role.

In the conservative free state, equation (10.1) becomes

$$[M]\,\ddot{\vec{x}} + [K]\,\vec{x} = \vec{0} \qquad (11.1)$$

Let us pre-multiply this equation by $[M]^{-1}$

$$\ddot{\vec{x}} + [M]^{-1}\,[K]\,\vec{x} = \vec{0} \qquad (11.2)$$

One calls the following product the **"core" of the system**

$$[A] = [M]^{-1}\,[K] \qquad (11.3)$$

With this notation, equation (11.2) can be written

$$\ddot{\vec{x}} + [A]\,\vec{x} = \vec{0} \qquad (11.4)$$

This matrix equation corresponds to a system of n second order differential equations which it is of interest to solve by two different methods

- a linear combination of particular solutions,
- a change of basis yielding the normal (or decoupled) coordinates.

## 11.2 Solution of the system by linear combination of specific solutions

### 11.2.1 Search for specific solutions

In the first way to tackle the problem, one initially searches for solutions $x_i(t)$ of the system (11.4) which are the same function of time, apart from a factor

$$x_i(t) = X_i\ g(t)$$

On the mathematical level, it is easy to show that $g(t)$ must be an harmonic function. The following physical reason suffices to convince one of this :

· no energy is supplied to the mechanical system for $t > 0$;
· energy cannot be dissipated by the system itself

Thus the total energy determined by the initial displacements and the initial velocities is conserved indefinitely.

For the vector of displacements, one therefore searches for a solution of the type

$$\vec{x} = \vec{X} \cos(\omega t-\varphi) \tag{11.5}$$

This gives, by introducing it into (11.4)

$$[- \omega^2 \lceil I \rfloor + [A]]\ \vec{X} \cos(\omega t-\varphi) = \vec{0}$$

By adopting the notation

$$\delta = \omega^2 \tag{11.6}$$

one obtains, after cancelling the $\cos(\omega t-\varphi)$ term

$$[[A] - \delta \lceil I \rfloor]\ \vec{X} = \vec{0} \tag{11.7}$$

This is an homogeneous system of equations which has non-zero solutions only if its determinant is null.

$$\left| [A] - \delta \lceil I \rfloor \right| = 0 \tag{11.8}$$

which gives, on expansion

$$\begin{vmatrix} (a_{11} - \delta) & a_{12} & \dots & a_{1n} \\ a_{21} & (a_{22} - \delta) & \dots & a_{2n} \\ \vdots & \vdots & & \vdots \\ a_{n1} & a_{n2} & \dots & (a_{nn} - \delta) \end{vmatrix} = 0 \tag{11.9}$$

The scalars $\delta = \omega^2$ are thus **the eigenvalues** of the core matrix [A]. The preceding equation, known as the **frequency equation** of the oscillating system or the **characteristic equation**, is of the form

$$\delta^n + \alpha_1 \delta^{n-1} + \alpha_2 \delta^{n-2} + \dots + \alpha_{n-1} \delta + \alpha_n = 0 \tag{11.10}$$

For physical reasons previously stated, it has solutions which are all positive (this assertion will be demonstrated in paragraph 11.3.3) which we shall assume, moreover, to be all distinct. By classing these solutions in increasing order one can write

$$\delta_1 < \delta_2 < \dots < \delta_p < \dots < \delta_n \tag{11.11}$$

which gives, on returning to the natural angular frequencies,

$$\omega_1 < \omega_2 < \dots < \omega_p < \dots < \omega_n \tag{11.12}$$

These angular frequencies of the conservative system could be designated $\omega_{01}, \dots, \omega_{0n}$ by analogy with the angular frequency $\omega_0$ of the conservative elementary oscillator. As we have done for the oscillator with two degrees of freedom, we shall not use this more detailed notation because confusion is not possible in the present chapter.

To each one of the natural angular frequencies $\omega_p$ corresponds a particular solution (11.5) of the differential system

$$\vec{x}_p = \vec{X}_p \cos(\omega_p t - \varphi_p) \tag{11.13}$$

or, using an indexed notation

$$\begin{cases} x_{1p} = X_{1p} \cos(\omega_p t - \varphi_p) \\ \vdots \\ x_{ip} = X_{ip} \cos(\omega_p t - \varphi_p) \\ \vdots \\ x_{np} = X_{np} \cos(\omega_p t - \varphi_p) \end{cases} \tag{11.14}$$

11.2.2 General solution · Natural modes

In order to avoid all confusion in what follows, it is necessary to remember that in all the expressions (11.14) the first index (i) designates a coordinate of the system while the second index (p) designates a natural angular frequency.

The vector $\vec{x}_p(t)$ represents the motion of the system at the natural angular frequency $\omega_p$ ; one calls it the pth **natural mode.** The vector $\vec{X}_p$ gives the amplitudes of $\vec{x}_p$ , one calls it the pth **natural mode shape.**

One obtains the components of $\vec{X}_p$ by introducing the value $\delta = \delta_p$ into the equations (11.7). It is thus necessary to solve $n$ systems of homogeneous equations successively. The components of $\vec{X}_p$ only being determined to a nearest factor, one can normalize the natural mode shapes in different ways, as we shall see later. One way to proceed, without interest on the theoretical level but convenient in practice, consists in taking the ratio of the amplitudes $X_{ip}$ to a reference amplitude $X_p$ (for example that of the first mass or that of the central mass in the case of geometrical symmetry of the system). One will write therefore

$$\beta_{ip} = \frac{X_{ip}}{X_p} \quad \Rightarrow \quad \vec{X}_p = \vec{\beta}_p X_p \tag{11.15}$$

The general solution of the differential system is given by a linear combination of the particular solutions, that is to say by (11.13)

$$\vec{x} = \sum_p^n \gamma_p \vec{x}_p = \sum_p^n \gamma_p \vec{\beta}_p X_p \cos(\omega_p t - \varphi_p)$$

As the amplitudes $X_p$ are arbitrary, one does not diminish the generality of the result by choosing $\gamma_p = 1$, whence finally

$$\vec{x} = \sum_p^n \vec{\beta}_p X_p \cos(\omega_p t - \varphi_p) \tag{11.16}$$

Let us make the expansion

$$\begin{Bmatrix} x_1 \\ \cdot \\ \cdot \\ \cdot \\ x_i \\ \cdot \\ \cdot \\ \cdot \\ x_n \end{Bmatrix} = \underbrace{\begin{Bmatrix} \beta_{11} \\ \cdot \\ \cdot \\ \cdot \\ \beta_{i1} \\ \cdot \\ \cdot \\ \cdot \\ \beta_{n1} \end{Bmatrix} X_1 \cos(\omega_1 t - \varphi_1)}_{\text{1st mode}} + \dots + \underbrace{\begin{Bmatrix} \beta_{1p} \\ \cdot \\ \cdot \\ \cdot \\ \beta_{ip} \\ \cdot \\ \cdot \\ \cdot \\ \beta_{np} \end{Bmatrix} X_p \cos(\omega_p t - \varphi_p)}_{\text{mode p}} + \dots \tag{11.17}$$

$$\dots + \underbrace{\begin{Bmatrix} \beta_{1n} \\ \cdot \\ \cdot \\ \cdot \\ \beta_{in} \\ \cdot \\ \cdot \\ \cdot \\ \beta_{nn} \end{Bmatrix} X_n \cos(\omega_n t - \varphi_n)}_{\text{mode n}}$$

### 11.2.3 Other forms of the characteristic equation

It is sometimes more convenient to calculate the influence coefficients than the stiffness of an elastic system. One can then, in principle, calculate the matrix $[K]$ by inverting the matrix $[\alpha]$. However, this operation is not very practical. It is more convenient to return to equation (11.1)

$$[M]\,\ddot{\vec{x}} + [K]\,\vec{x} = \vec{0}$$

and pre-multiply by $[K]^{-1} = [\alpha]$

$$[\alpha]\,[M]\,\ddot{\vec{x}} + \vec{x} = \vec{0} \tag{11.18}$$

The product $[\alpha]\,[M]$ is the inverse of the previously defined core.

Let us write

$$[E] = [\alpha]\,[M] = [A]^{-1} \tag{11.19}$$

It gives therefore

$$[E]\,\ddot{\vec{x}} + \vec{x} = \vec{0} \tag{11.20}$$

By seeking the harmonic solutions

$$\vec{x} = \vec{X}\cos(\omega t-\varphi) = \vec{X}\cos(\sqrt{\delta}\,t-\varphi)$$

one finds the characteristic equation

$$\left| -\delta[E] + \lceil I \rfloor \right| = 0 \tag{11.21}$$

whose solutions are naturally the same as those of (11.8). One readily gives this equation the following form

$$\left| [E] - \tau\lceil I \rfloor \right| = 0 \tag{11.22}$$

in which $\tau$ is called the **frequency function**

$$\tau = \frac{1}{\delta} \tag{11.23}$$

It is necessary to point out that the characteristic equations (11.8) and (11.22) are not suitable for numerical calculations. Indeed, the matrices [A] and [E] are not symmetrical whereas the matrices [M] and [K] are. It is more convenient, for such calculations, to choose the characteristic equation from one of the forms

$$\left| [K] - \delta [M] \right| = 0 \tag{11.24}$$

$$\left| [\alpha] - \tau [M]^{-1} \right| = 0 \tag{11.25}$$

which one deduces immediately from the preceding equations.

### 11.2.4 Summary and comments · Additional constraints

It is useful, already at the present stage of this study, to recall the principal results obtained and to make several comments. In particular, we are going to point out the role played by one or more additional constraints.

- An oscillating mechanical (or electrical, electromagnetic, or other system) with $n$ degrees of freedom possesses $n$ natural angular frequencies. We shall assume for the moment that all these angular frequencies are distinct.

- For each of the natural angular frequencies $\omega_p$ there correspond the synchronous harmonic displacements $x_{ip}(t)$ , which appear with different amplitudes in the total displacements $x_i(t)$ . These displacements make up together the pth **natural mode.** The vector $\vec{x}(t)$ of the total displacements of the system is therefore the sum of the natural modes, as equations (11.16) or (11.17) show.

- These equations have 2n constants of integration, that is to say n amplitudes $X_p$ and n phases $\varphi_p$ , which are determined by the initial conditions. One can thus choose the initial conditions, at least in principle, in such a way that there exists only one single mode at a time, all the others being zero.

- by analogy with continuous vibrating systems which possess an infinite number of natural angular frequencies, the lowest of the natural angular frequencies and the corresponding mode are called the **fundamental angular frequency** and the **fundamental mode**.

- So we assume that the oscillating system studied has n degrees of freedom and n natural angular frequencies arranged in ascending order as follows

$$\omega_1 < \omega_2 < \ldots < \omega_n$$

If one adds a constraint, the number of degrees of freedom is reduced to (n-1) . One can show that the (n-1) new natural angular frequencies $\omega_1'$ , $\omega_2'$ ... $\omega'_{n-1}$ are included between the preceding ones

$$\omega_1 < \omega_1' < \omega_2 < \omega_2' < \omega_3 \ldots < \omega_{n-1} < \omega_{n-1}' < \omega_n \qquad (11.26)$$

- In a more general way, if one adds k constraints, the natural angular frequency $\omega_p'$ of the modified system is included between the angular frequencies $\omega_p$ and $\omega_{p+k}$ of the initial system,

$$\omega_p < \omega_p' < \omega_{p+k} \qquad (11.27)$$

## 11.3 Solution of the system by change of coordinates

### 11.3.1. Decoupling of the equations · Normal coordinates

The general solution of the differential system (11.1) has been obtained by linear superposition of specific solutions, called **natural modes of the system.** In order to establish other properties of the solutions, which are of great importance in the methods of analysis of the vibrations, it is necessary to adopt a different approach.

Let us again use the differential equation in the matrix form (11.1)

$$[M]\,\ddot{\vec{x}} + [K]\,\vec{x} = \vec{0}$$

This represents a system of $n$ second order differential equations with constant coupled coefficients. That means that the ith equation, for example, can be a function of all the variables $x_j$ and their derivatives $\ddot{x}_j$ , with $j = 1, 2, \ldots, n$ .

The principle of the method which we are going to develope consists of seeking a new set of generalized coordinates $q_p$ , $p = 1, \ldots, n$ so that the motion of the system is described by $n$ decoupled differential equations. These new coordinates, functions only of time, will be linear combinations of the initial coordinates $x_i$. Each equation $p$ will then depend only on the variable $q_p$ and its second derivative $\ddot{q}_p$ . In this way we will be led to consider n systems analogous to an elementary oscillator. (11.28)

$$m_p^\circ\,\ddot{q}_p + k_p^\circ\,q_p = 0 \qquad p = 1, 2, \ldots, n$$

or, by dividing by $m_p^0$

$$\ddot{q}_p + \delta_p \, q_p = 0 \qquad p = 1, 2, \ldots, n \tag{11.29}$$

with

$$\delta_p = \frac{k_p^0}{m_p^0} \tag{11.30}$$

The coordinates $q_p$, called **normal coordinates** (or **modal coordinates)** of the system, are linear combinations of the $x_i$ and vice versa. It is therefore possible to make the following change of variables

$$\vec{x} = [B] \, \vec{q} \tag{11.31}$$

The **matrix** [B], **known** as the **change of basis matrix,** is square, regular and of dimension n. Its coefficients are independent of time. The accelerations are then linked by the obvious relation

$$\ddot{\vec{x}} = [B] \, \ddot{\vec{q}} \tag{11.32}$$

By making the change of variables in the equation (11.1), we get

$$[M] \, [B] \, \ddot{\vec{q}} + [K] \, [B] \, \vec{q} = \vec{0} \tag{11.33}$$

The matrices [M] and [K] being symmetrical, we can pre-multiply this relationship by $[B]^T$ in order to re-find the symmetry

$$[B]^T \, [M] \, [B] \, \ddot{\vec{q}} + [B]^T \, [K] \, [B] \, \vec{q} = \vec{0} \tag{11.34}$$

This differential system corresponds to n decoupled equations, of the form (11.28), if the symmetric matrices $[B]^T[M] \, [B]$ and $[B]^T[K] \, [B]$ are simultaneously diagonal. It is then a matter of determining if there exists a matrix [B] - and the matrix itself - which satisfies the conditions :

$$\begin{cases} [M\rfloor = [B]^T \, [M] \, [B] \\ [K\rfloor = [B]^T \, [K] \, [B] \end{cases} \tag{11.35}$$

Without demonstrating it here, we can maintain, according to the spectral theorem [22] , that there exists an infinite number of invertible matrices [B] , of the same natural directions, such that $[B]^T[M]\ [B]$ and $[B]^T[K]\ [B]$ are simultaneously diagonal and real. It is necessary and sufficient for this that matrices [M] and [K] are both symmetrical and that one of them, at least, is positive definite. Such is actually the case, as we have shown in chapter 10.

Let us re-write the differential system (11.34) by pre-multiplying it by $[[B]^T[M]\ [B]]^{-1}$. One obtains

$$\ddot{\vec{q}} + [B]^{-1}[M]^{-1}[K]\ [B]\ \vec{q} = \vec{0} \tag{11.36}$$

By comparing this relation to (11.29), where the terms $\delta_p$ are the elements of the diagonal matrix $[\Delta]$, one can write

$$[B]^{-1}[M]^{-1}[K]\ [B] = \lceil\Delta\rfloor \tag{11.37}$$

The matrix sought [B] is thus that which allows one to diagonalize the core $[A] = [M]^{-1}[K]$ of the system

$$[B]^{-1}[A]\ [B] = \lceil\Delta\rfloor \tag{11.38}$$

The diagonalization of a square matrix is a standard operation of linear algebra. So, even before proceeding to the calculations which are going to follow, we know already that the terms of the diagonal matrix $\lceil\Delta\rfloor$ are the eigenvalues of the matrix [A] and that the columns of the matrix [B] are made up of the eigenvectors of [A] corresponding to each eigenvalue.

### 11.3.2 Eigenvalue problem

Let us re-write (11.38) by multiplying the two sides by [B]

$$[A]\ [B] = [B]\ \lceil\Delta\rfloor \tag{11.39}$$

By designating the columns of the matrix [B] by $\vec{B}_p$ , this relation takes the form

$$[[A]\ \vec{B}_1\ |\ldots|\ [A]\ \vec{B}_p\ |\ldots|\ [A]\ \vec{B}_n] = [\delta_1\ \vec{B}_1\ |\ldots|\ \delta_p \vec{B}_p\ |\ldots|\ \delta_n\ \vec{B}_n] \tag{11.40}$$

The matrix of the left-hand side and that of the right-hand side are equal if their columns are identical, in other words

$$[A]\ \vec{B}_p = \delta_p\ \vec{B}_p \tag{11.41}$$

Since $\delta_p$ is a number, this result can be written

$$[[A] - \delta_p \lceil I \rfloor]\ \vec{B}_p = \vec{0} \tag{11.42}$$

The system (11.42) has n equations with in n unknowns. It only has non-zero solutions if its determinant is null.

$$\left| [A] - \delta_p \lceil I \rfloor \right| = 0$$

One has found again the characteristic equation (11.8) allowing one to calculate the eigenvalues of [A] , which we have previously assumed to be all distinct. Once these eigenvalues are obtained, equation (11.42) enables one to calculate the columns of the matrix [B] apart from a scale factor. It is thus a matter of solving successively n homogeneous systems of n algebraic equations with n unknowns.

Let us point out that the eigenvector $\vec{B}_p$ is proportional to the column vectors, of the adjoint matrix of $[[A] - \delta_p \lceil I \rfloor]$.(These column vectors are equal apart from a scale factor.)

We have thus shown that it is always possible to transform the differential system (11.1) into a decoupled system

$$\lceil M \rfloor\ \ddot{\vec{q}} + \lceil K \rfloor\ \vec{q} = \vec{0} \tag{11.43}$$

Let us recall that $\vec{x} = [B]\,\vec{q}$ and that matrices $\lceil M \rfloor$ and $\lceil K \rfloor$ are defined by (11.35). By dividing (11.43) by $\lceil M \rfloor$, one obtains

$$\ddot{\vec{q}} + \lceil M \rfloor^{-1} \lceil K \rfloor\,\vec{q} = \vec{0} \tag{11.44}$$

By taking account of the definition (11.37), the comparison of this result with (11.36) shows that

$$\lceil M \rfloor^{-1} \lceil K \rfloor = \lceil \Delta \rfloor \tag{11.45}$$

This gives finally

$$\ddot{\vec{q}} + \lceil \Delta \rfloor\,\vec{q} = \vec{0} \tag{11.46}$$

which is the system of $n$ **decoupled second order differential equations** (11.29) which we are trying to obtain.

### 11.3.3 Energetic forms - Sign of the eigenvalues

One knows, from (10.3) and (10.2) that the potential and kinetic energies have the values

$$V = \frac{1}{2}\,\vec{x}^T [K]\,\vec{x}$$

$$T = \frac{1}{2}\,\dot{\vec{x}}^T [M]\,\dot{\vec{x}}$$

The change of coordinates $\vec{x} = [B]\,\vec{q}$ gives thus

$$V = \frac{1}{2}\,\vec{q}^T [B]^T [K]\,[B]\,\vec{q}$$

$$T = \frac{1}{2}\,\dot{\vec{q}}^T [B]^T [M]\,[B]\,\dot{\vec{q}}$$

which gives, taking account of the relations (11.35) :

$$V = \frac{1}{2}\, \vec{q}^T [K] \, \vec{q} = \frac{1}{2} \sum_p^n k_p^0 \, q_p^2 \tag{11.47}$$

$$T = \frac{1}{2}\, \dot{\vec{q}}^T [M] \, \dot{\vec{q}} = \frac{1}{2} \sum_p^n m_p^0 \, \dot{q}_p^2 \tag{11.48}$$

Thus, as one had to expect, the kinetic and potential energies, expressed in the modal basis, are the sums of n independent quantities.

We are going to show that the **eigenvalues of the system are all positive or zero.** The terms $m_p^0$ of the matrix $[M]$ are all exclusively positive since the kinetic energy is a positive definite quadratic form.

$$m_p^0 > 0 \tag{11.49}$$

The terms $k_p^0$ of $[K]$ can be positive or zero for the potential energy is a positive semi-definite quadratic form.

$$k_p^0 \geq 0 \tag{11.50}$$

It follows that

$$\delta_p = \frac{k_p^0}{m_p^0} \geq 0 \tag{11.51}$$

which justifies writing

$$\delta_p = \omega_p^2 \tag{11.52}$$

### 11.3.4 General form of the solution

Taking into account (11.52), integration of equations (11.29) is immediate

$$q_p = Q_p \cos(\omega_p t - \varphi_p) \qquad p = 1, 2, \ldots, n \tag{11.53}$$

Let us return to the initial coordinates of the system by means of (11.31)

$$\vec{x} = [B]\,\vec{q}$$

one obtains

$$\vec{x} = \sum_p^n \vec{B}_p\, q_p \tag{11.54}$$

and finally, by replacing the $q_p$ by their expressions above

$$\vec{x} = \sum_p^n \vec{B}_p\, Q_p \cos(\omega_p t - \varphi_p) \tag{11.55}$$

The normalization of $\vec{B}_p$ being undefined, the change of notation $\vec{B}_p\, Q_p = \vec{\beta}_p\, X_p$ enables one to find (11.16) again

$$\vec{x} = \sum_p^n \vec{\beta}_p\, X_p \cos(\omega_p t - \varphi_p)$$

We note thus that the quantities $q_p$ , which we have called **normal coordinates, correspond to the natural modes of the system.**

The eigenvectors of the core [A] are proportional to the columns $\vec{B}_p$ of the matrix [B] , so they are also proportional to the vectors $\vec{\beta}_p$ appearing in the previous relationship. These are the **modal vectors,** called the **natural mode shapes** of the system. They are the static deformations or displacements associated with each natural angular frequency.

One knows that every vector proportional to an eigenvector is itself an eigenvector. Consequently, in what follows, the symbols $\vec{B}_p$ and $\vec{\beta}_p$ are equivalent. The choice will depend only on the ease of writing in order to express the result sought.

Since the potential energy of the system is positive semi-definite, it can eventually cancel itself out without the $q_p$ being all simultaneously zero. This means then that at least one of the $k^0$ is zero, as well as the corresponding angular frequency. A movement $q_p$ at zero angular frequency is called **a zero mode** or **rigid body mode.** A system having this characteristic is described as semi-definite. If a system possesses $r$ zero modes, the rank of the matrices $[K]$ , $\lceil K \rfloor$ and $[A]$ is then $n - r$ .

### 11.3.5 Linear independence and orthogonality of the modal vectors

The directions of the linearly independent coordinates $x_i$ , constitute a basis for the **configuration space**. In the same way, the modal vectors are linearly independent and form a new basis for this space. This basis and the matrix $[B]$ are called the **modal basis** and the **modal matrix** respectively.

The linear independence of the modal vectors is expressed by the relation

$$\sum_p^n \gamma_p \vec{\beta}_p \neq \vec{0} \qquad (11.56)$$

in which the $\gamma_p$ are arbitrary constants, not all of which are zero. This is easily demonstrated (but we will not do that here) from another property of vectors known as the **orthogonality of modal vectors** or **orthogonality of modes**. The relations expressing this last property, which is fundamental in modal analysis, can be established from expressions (11.35). Let us start with the first

$$[B]^T[M]\ [B] = [M^0]$$

By introducing the modal vectors $\vec{\beta}_p$ one can write

$$\begin{bmatrix} \vec{\beta}_1^T \\ \vdots \\ \vec{\beta}_n^T \end{bmatrix} [M]\ [\ \vec{\beta}_1\ \ldots\ \vec{\beta}_n] = \begin{bmatrix} m_1^0 & & 0 \\ & \ddots & \\ 0 & & m_n^0 \end{bmatrix} \tag{11.57}$$

By means of identification term by term, one obtains

$$\vec{\beta}_r^T[M]\ \vec{\beta}_r = m_r^0 \tag{11.58}$$

$$\vec{\beta}_r^T[M]\ \vec{\beta}_s = 0 \qquad r \neq s \tag{11.59}$$

which gives, using the Kronecker delta symbol $\delta_{rs}$

$$\vec{\beta}_r^T[M]\ \vec{\beta}_s = \delta_{rs}\ m_r^0 \tag{11.60}$$

The relation (11.59) expresses the **orthogonality of the modal vectors** whereas the relation (11.58) defines the **modal mass** $m_r^0$ attributed to the rth mode of the system. The effective value of $m_r^0$ depends on the choice made for the normalization of the vectors $\vec{\beta}_p$ .

By repeating the calculation using the second relation in (11.25), one obtains in the same way

$$\vec{\beta}_r^T[K]\ \vec{\beta}_s = \delta_{rs}\ k_r^0 \tag{11.61}$$

In this relation, $k_r^0$ is the **modal stiffness** attributed to the rth mode.

**The orthogonality of the modal vectors can then be summarized thus : the scalar product, weighted by the mass matrix or by the stiffness matrix, of two natural mode shapes or natural modes of different order is zero.**

When the mass matrix is diagonal with the terms all equal, $[M] = m_0[I]$ , the orthogonality takes the form of a direct scalar product

$$\vec{\beta}_r^T \cdot \vec{\beta}_s = 0 \qquad r \neq s \tag{11.62}$$

11.3.6 Normalization of natural mode shapes

The fact that the modal vectors making up the modal matrix [B] are defined using an arbitrary scale factor justifies any choice of normalization for the natural modes shapes. Here are some examples.

- One gives a unit value to the amplitude of a specific variable $i=m$ , whatever the order of the mode : $X_{mp} = 1$ . In general, it is the first variable, so that $X_{1p} = 1$ .

- One gives a unit value to the largest of the amplitudes appearing in a mode, $(X_{ip})_{max} = 1$ .

- The modal masses are all set equal to one, which is to say

$$[B]^T[M]\,[B] = \lceil I \rfloor \tag{11.63}$$

  This normalization is adopted in problems of identification of structures, of sensitivity analysis, etc.

- A unit length is attributed to each modal vector, so that

$$\|\vec{\beta}_p\| = \sum_i^n \beta_{ip}^2 = 1 \tag{11.64}$$

  This is a form of normalization often adopted for the graphical representation of natural mode shapes.

## 11.4 Response to an initial excitation

Let us return to the solutions (11.16) of the system

$$\vec{x} = \sum_p^n \vec{\beta}_p X_p \cos(\omega_p t - \varphi_p)$$

$$= \sum_p^n \vec{\beta}_p X_p (\cos \varphi_p \cos \omega_p t + \sin \varphi_p \sin \omega_p t)$$

and let us consider a release with any initial conditions

$$\vec{x}(0) = \vec{X}_0 \tag{11.65}$$

$$\dot{\vec{x}}(0) = \vec{V}_0 \tag{11.66}$$

This gives successively

$$\vec{X}_0 = \sum_p^n \vec{\beta}_p X_p \cos \varphi_p \tag{11.67}$$

$$\vec{V}_0 = \sum_p^n \vec{\beta}_p \omega_p X_p \sin \varphi_p \tag{11.68}$$

In order to eliminate the constants $X_p$ and $\varphi_p$ , let us pre-multiply the relations above by $\vec{\beta}_r^T[M]$ ,

$$\vec{\beta}_r^T[M] \vec{X}_0 = \sum_p^n \vec{\beta}_r^T[M] \vec{\beta}_p X_p \cos \varphi_p$$

$$\vec{\beta}_r^T[M] \vec{V}_0 = \sum_p^n \vec{\beta}_r^T[M] \vec{\beta}_p \omega_p X_p \sin \varphi_p$$

By using the orthogonality relations (11.60), one ends up with the equations

$$X_r \cos \varphi_r = \frac{1}{m_r^0} \vec{\beta}_r^T [M] \vec{X}_0 \qquad (11.69)$$

$$X_r \sin \varphi_r = \frac{1}{m_r^0 \, \omega_r} \vec{\beta}_r^T [M] \vec{V}_0 \qquad (11.70)$$

The response of the system to the initial conditions above is thus

$$\vec{x} = \sum_p^n \frac{1}{m_p^0} \vec{\beta}_p (\vec{\beta}_p^T [M] \vec{X}_0 \cos \omega_p t + \frac{1}{\omega_p} \vec{\beta}_p^T [M] \vec{V}_0 \sin \omega_p t) \qquad (11.71)$$

Let us examine the particular case where the vector of the initial displacements is proportional to one of the eigenvectors $\vec{\beta}_r$ and where the vector of initial velocities is null

$$\vec{X}_0 = X_0 \vec{\beta}_r \qquad (11.72)$$

$$\vec{V}_0 = \vec{0} \qquad (11.73)$$

The response of the system is then

$$\vec{x} = X_0 \sum_p^n \frac{1}{m_p^0} \vec{\beta}_p (\vec{\beta}_p^T [M] \vec{\beta}_r) \cos \omega_p t$$

and finally, because of the orthogonality of the modes (11.60)

$$\vec{x} = X_0 \vec{\beta}_r \cos \omega_r t \qquad (11.74)$$

This result shows that the system vibrates only in the r th mode . In this way, one can isolate a vibratory mode, in a conservative system with n degrees of freedom in the free state, by choosing the initial conditions (11.72) and (11.73).

If one replaces the condition of zero initial velocities by the less restrictive condition

$$\vec{V}_0 = V_0 \vec{\beta}_r \tag{11.75}$$

the motion of the system also occurs only in the rth mode. In effect, one easily finds in this case

$$\vec{x} = \sqrt{X_0^2 + (\frac{V_0}{\omega_r})^2}\ \vec{\beta}_r \cos(\omega_r t - \varphi_r) \qquad \mathrm{tg}\ \varphi_r = \frac{V_0}{X_0\ \omega_r} \tag{11.76}$$

## 11.5 Rayleigh quotient

By replacing the core matrix [A] by its value $[M]^{-1}[K]$ , the eigenvalue problem (11.42), examined previously, can be put in the form

$$\delta_p[M]\ \vec{\beta}_p = [K]\ \vec{\beta}_p\ (\delta_p = \omega_p^2) \tag{11.77}$$

In this case, when knowledge of all the modes and natural frequencies is not required and one searches only for an estimate of the fundamental angular frequency, use of the Rayleigh quotient turns out to be very convenient. In order to introduce this concept, let us consider the solutions $\delta_p$ , $\vec{\beta}_p$ (p = 1, 2, ..., n) satisfying

$$\delta_p[M]\ \vec{\beta}_p = [K]\ \vec{\beta}_p \qquad p = 1, 2, ..., n$$

By pre-multiplying the two sides by $\vec{\beta}_p^T$ , then by dividing by the scalar $\vec{\beta}_p^T[M]\ \vec{\beta}_p$ , that is to say by the modal mass $m_p^0$ , one obtains

$$\delta_p = \omega_p^2 = \frac{\vec{\beta}_p^T[K]\ \vec{\beta}_p}{\vec{\beta}_p^T[M]\ \vec{\beta}_p} \qquad p = 1, 2, \ldots, n \tag{11.78}$$

In this way the eigenvalues can be written in the form of the quotient of two triple products of matrices representing the quadratic forms; the numerator corresponds to the potential energy and the denominator to the kinetic energy of the mode considered. Let us write this same ratio for any vector $\vec{u}$ . It gives

$$\delta = \omega^2 = R(u) = \frac{\vec{u}^T[K]\ \vec{u}}{\vec{u}^T[M]\ \vec{u}} \tag{11.79}$$

One calls the scalar quantity $R(u)$ defined in this way the **Rayleigh quotient,** which depends on the matrices $[K]$ and $[M]$ as well as $\vec{u}$ .

For a given system the Rayleigh quotient depends only on $\vec{u}$ , the matrices $[M]$ and $[K]$ being determined by the characteristics of the system. This quotient posseses interesting properties that we are going to establish below.

Let us express any vector $\vec{u}$ as a linear combination of the modal vectors (11.56)

$$\vec{u} = \sum_p^n \gamma_p\ \vec{\beta}_p \tag{11.80}$$

or again, in matrix form, $[B]$ being the modal matrix and $\vec{\gamma}$ the vector made up from the coefficients $\gamma_p$

$$\vec{u} = [B]\ \vec{\gamma} \tag{11.81}$$

On the other hand, let us assume that the modal vectors are normalized in such a way that

$$[B]^T[M]\ [B] = \lceil I \rfloor$$

This gives, taking into account (11.35) and (11.45) :

$$[B]^T[K]\ [B] = \lceil \Delta \rfloor$$

One can then introduce the relation (11.81) into the definition (11.79) for the Rayleigh quotient and then use the above two relations

$$R(u) = \frac{\vec{\gamma}^T [B]^T [K] \; [B] \; \vec{\gamma}}{\vec{\gamma}^T [B]^T [M] \; [B] \; \vec{\gamma}} = \frac{\vec{\gamma}^T \lceil \Delta \rfloor \; \vec{\gamma}}{\vec{\gamma}^T \lceil I \rfloor \; \vec{\gamma}} \qquad (11.82)$$

In this last result, the matrix products are the sums of $n$ scalar quantities

$$R(u) = \frac{\sum_p^n \delta_p \; \gamma_p^2}{\sum_p^n \gamma_p^2} \qquad (11.83)$$

When the vector $\vec{u}$ is not very different from the eigenvector $\vec{\beta}_r$, the $\vec{\gamma}_p$ for $p \neq r$, are very small compared to $\gamma_r$.

$$\gamma_p = \varepsilon_p \; \gamma_r \qquad p = 1, 2, \ldots, n \qquad (\varepsilon_p \ll 1 \; \forall \; p \neq r) \qquad (11.84)$$

By introducing (11.84) into (11.83) and by dividing by $\gamma_r^2$ one obtains

$$R(u) = \frac{\delta_r + \sum_{p \neq r}^n \delta_p \; \varepsilon_p^2}{1 + \sum_{p \neq r}^n \varepsilon_p^2} \qquad (11.85)$$

Let us calculate the approximate value of R(u) in the neighbourhood of $\varepsilon_p = 0$

$$R(u) \approx (\delta_r + \sum_{p \neq r}^n \delta_p \; \varepsilon_p^2) \; (1 - \sum_{p \neq r}^n \varepsilon_p^2)$$

$$R(u) \approx \delta_r + \sum_{p \neq r}^n (\delta_p - \delta_r) \; \varepsilon_p^2 \qquad (11.86)$$

When the vector $\vec{u}$ only differs from $\vec{\beta}_r$ by a small first order quantity, R(u) only differs from $\delta_r = \omega_r^2$ by a small second order quantity. Thus the Rayleigh quotient R(u) has a stationary value in the neighbourhood of an eigenvector.

However, the most important property of the Rayleigh quotient is that it presents a minimun in the vicinity of the fundamental mode.

In effect, if $r = 1$ in (11.86), one obtains

$$R(u) \simeq \delta_1 + \sum_{p=2}^{n} (\delta_p - \delta_1)\, \varepsilon_p^2 \qquad (11.87)$$

The eigenvalues are ordered as follows

$$\delta_i > \delta_1 \qquad (i = 2, \ldots, n)$$

As a result

$$R(u) \geqslant \delta_1 \qquad (11.88)$$

The Rayleigh quotient is thus always greater than or equal to the fundamental eigenvalue. The main use for the Rayleigh quotient is to make an estimate of the fundamental natural angular frequency of a system by using an approximate fundamental eigenvector, determined by physical considerations (an example is given in paragraph 11.6.3).

Finally, it is easy to verify that the relationship (11.79) can be put into the following equivalent form

$$R(u) = \frac{\vec{u}^T \vec{u}}{\vec{u}^T [\alpha]\, [M]\, \vec{u}} \qquad (11.89)$$

This is preferable when one knows the flexibility matrix $[\alpha]$.

## 11.6 Examples of generalized conservative oscillators

### 11.6.1 Symmetrical triple pendulum

The calculation of small movements of a system consisting of three massive, equal and symmetrically coupled pendulums (symmetrical triple pendulum, figure 11.1) is an elementary and classical example to make the concept of natural modes more concrete.

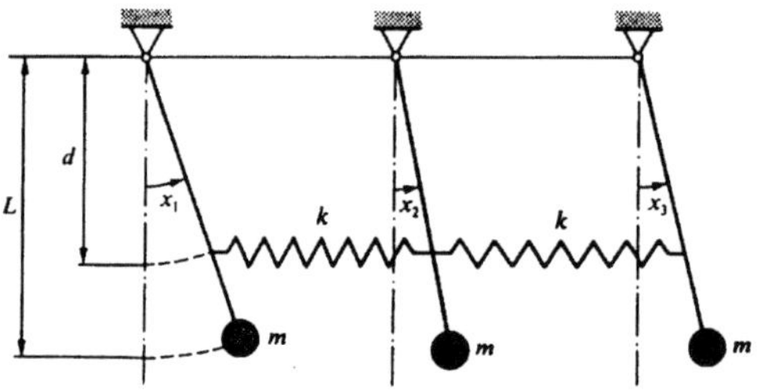


Fig. 11.1 Symmetrical triple pendulum

The kinetic energy has the value

$$T = \frac{1}{2} m \sum_{i}^{3} (L \dot{x}_i)^2 = \frac{1}{2} m L^2 (\dot{x}_1^2 + \dot{x}_2^2 + \dot{x}_3^2)$$

The potential energy is made up of two parts, that due to the elevation of the masses in the field of gravitation and that which corresponds to the deformation of the springs

$$V = m g L \sum_{i}^{3} (1 - \cos x_i)$$

$$+ \frac{1}{2} k [(d \sin x_1 - d \sin x_2)^2 + (d \sin x_2 - d \sin x_3)^2]$$

Using Lagrange's equations

$$\frac{d}{dt}\left(\frac{\partial T}{\partial \dot{x}_i}\right) + \frac{\partial V}{\partial x_i} = 0 \qquad i = 1,\ 2,\ 3$$

one differentiates and then adopts the small angle hypothesis ($\sin x \approx x$ and $\cos x \approx 1$). We get in this way :

$$\begin{cases} \ddot{x}_1\, m\, L^2 + x_1(m\, g\, L + k\, d^2) - x_2\, k\, d^2 = 0 \\ \ddot{x}_2\, m\, L^2 - x_1\, k\, d^2 + x_2(m\, g\, L + 2\, k\, d^2) - x_3\, k\, d^2 = 0 \\ \ddot{x}_3\, m\, L^2 - x_2\, k\, d^2 + x_3(m\, g\, L + k\, d^2) = 0 \end{cases} \tag{11.90}$$

By comparing these equations with the matrix equation

$$[M]\,\ddot{\vec{x}} + [K]\,\vec{x} = \vec{0}$$

And by defining the ratio, proportional to the coupling

$$\mu = \frac{k\, d^2}{m\, g\, L} \tag{11.91}$$

one sees that the mass matrix and the stiffness matrix have the values

$$[M] = m\, L^2 [I] \qquad [K] = m\, g\, L \begin{bmatrix} 1+\mu & -\mu & 0 \\ -\mu & 1+2\,\mu & -\mu \\ 0 & -\mu & 1+\mu \end{bmatrix} \tag{11.92}$$

The core of the system is particularly simple to calculate here

$$[A] = [M]^{-1}[K] = \frac{g}{L} \begin{bmatrix} 1+\mu & -\mu & 0 \\ -\mu & 1+2\,\mu & -\mu \\ 0 & -\mu & 1+\mu \end{bmatrix} \tag{11.93}$$

Let us now try to find the eigenvalues $\delta$ from the characteristic equation (11.8)

$$\left|[A] - \delta[I]\right| = 0 \quad \Rightarrow \quad \left|\frac{L}{g}[A] - \frac{L\delta}{g}[I]\right| = 0$$

With the convention of writing

$$y = \frac{L}{g}\,\delta$$

This equation becomes

$$\begin{vmatrix} 1+\mu-y & -\mu & 0 \\ -\mu & 1+2\mu-y & -\mu \\ 0 & -\mu & 1+\mu-y \end{vmatrix} = 0$$

which gives, after expansion

$$(1-y)(1+\mu-y)(1+3\mu-y) = 0$$

The solutions give the values and natural angular frequencies

$$\begin{cases} y_1 = 1 \\ y_2 = 1+\mu \\ y_3 = 1+3\mu \end{cases} \Rightarrow \begin{cases} \delta_1 = \frac{g}{L} \\ \delta_2 = \frac{g}{L}(1+\mu) \\ \delta_3 = \frac{g}{L}(1+3\mu) \end{cases} \Rightarrow \begin{cases} \omega_1 = \sqrt{\frac{g}{L}} \\ \omega_2 = \omega_1\sqrt{1+\mu} \\ \omega_3 = \omega_1\sqrt{1+3\mu} \end{cases}$$

One notices already that the fundamental angular frequency $\omega_1$ is that of the simple pendulum. The coupling not taking place, the three pendulums must oscillate together, which goes to confirm the calculation of the natural mode shapes below.

Let us recall that this calculation requires the solution of the three homogeneous systems obtained by replacing $\delta$ by $\delta_1$, $\delta_2$ and $\delta_3$ successively.

One will naturally obtain the same results, apart from a normalization factor, by replacing $y$ by $y_1$, $y_2$ and $y_3$ in the terms of the characteristic equation. These terms are the coefficients of the equations to solve. As for the variables of these equations, they are proportional to the amplitudes of the oscillations. We will designate them by $r$, $s$, $t$ to simplify the notation.

First natural mode shape

$$y = y_1 = 1$$

$$\begin{cases} \mu r - \mu s = 0 \\ -\mu r + 2\mu s - \mu t = 0 \\ -\mu s + \mu t = 0 \end{cases} \Rightarrow \begin{cases} s = r \\ t = r \end{cases} \Rightarrow \vec{\beta}_1 = \begin{Bmatrix} 1 \\ 1 \\ 1 \end{Bmatrix} \quad (11.97)$$

Second natural mode shape

$$y = y_2 = 1 + \mu$$

$$\begin{cases} -\mu s = 0 \\ -\mu r + \mu s - \mu t = 0 \\ -\mu s = 0 \end{cases} \Rightarrow \begin{cases} s = 0 \\ t = -r \end{cases} \Rightarrow \vec{\beta}_2 = \begin{Bmatrix} 1 \\ 0 \\ -1 \end{Bmatrix} \quad (11.98)$$

Third natural mode shape

$$y = y_3 = 1 + 3\mu$$

$$\begin{cases} -2\mu r - \mu s = 0 \\ -\mu r - \mu s - \mu t = 0 \\ -\mu s - 2\mu t = 0 \end{cases} \Rightarrow \begin{cases} s = -2r \\ t = r \end{cases} \Rightarrow \vec{\beta}_3 = \begin{Bmatrix} 1 \\ -2 \\ 1 \end{Bmatrix} \quad (11.99)$$

Let us verify the orthogonality of the natural mode shapes, by using the relation (11.62) since $[M] = m L^2 [I]$

$$\vec{\beta}_1^T \cdot \vec{\beta}_2 = (1\ 1\ 1) \begin{Bmatrix} 1 \\ 0 \\ -1 \end{Bmatrix} = 1 + 0 - 1 = 0$$

$$\vec{\beta}_2^T \cdot \vec{\beta}_3 = (1\ 0\ -1) \begin{Bmatrix} 1 \\ -2 \\ 1 \end{Bmatrix} = 1 + 0 - 1 = 0$$

$$\vec{\beta}_3^T \cdot \vec{\beta}_1 = (1\ -2\ 1) \begin{Bmatrix} 1 \\ 1 \\ 1 \end{Bmatrix} = 1 - 2 + 1 = 0$$

Apart from the initial conditions, the free state is now entirely determined

$$\begin{Bmatrix} x_1 \\ x_2 \\ x_3 \end{Bmatrix} = \begin{Bmatrix} 1 \\ 1 \\ 1 \end{Bmatrix} X_1 \cos(\omega_1 t - \varphi_1) + \begin{Bmatrix} 1 \\ 0 \\ -1 \end{Bmatrix} X_2 \cos(\omega_2 t - \varphi_2) + \begin{Bmatrix} 1 \\ -2 \\ 1 \end{Bmatrix} X_3 \cos(\omega_3 t - \varphi_3) \quad (11.100)$$

The natural mode shapesare shown in figure 11.2.

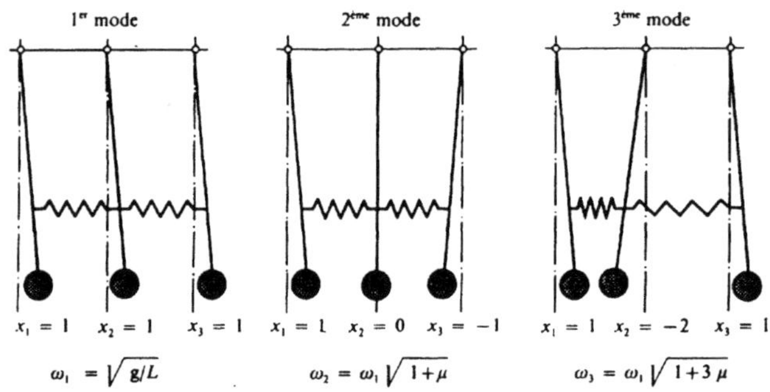


Fig. 11.2 Natural mode shapes and angular frequencies of the system of fig. 11.1

So as to show the influence of coupling on the natural frequencies, let us choose a numerical example,

$$L = 1 \text{ m} \qquad d = 0{,}5 \text{ m} \qquad m = 1 \text{ kg}$$

The fundamental frequency is thus

$$f_1 = \frac{1}{2\pi}\,\omega_1 = \frac{1}{2\pi}\sqrt{\frac{g}{L}} = 0.5 \text{ Hz}$$

By taking the stiffness of the springs as a variable, the coefficient $\mu$, which is proportional to the coupling, and the two other natural frequencies have the following values (figure 11.3)

$$\mu = \frac{d^2}{m\,g\,L}\,k = 0.0255\,k$$

$$f_2 = f_1\sqrt{1+\mu} \qquad\qquad f_3 = f_1\sqrt{1+3\mu}$$

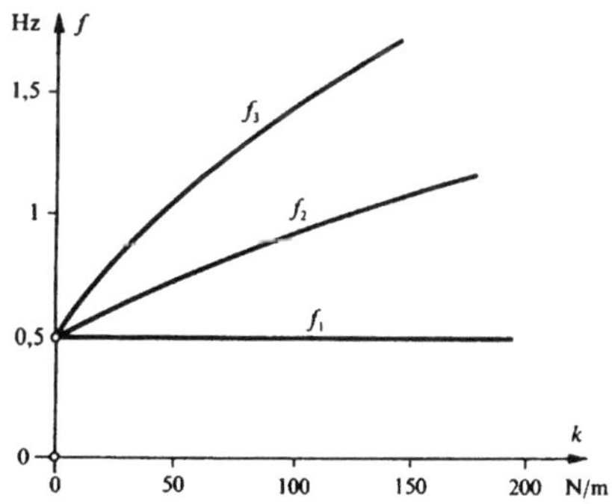


Fig. 11.3 Natural frequencies as a function of the stiffness of the coupling springs

The three natural frequencies become equal when the coupling tends to zero, as is to be expected. The system then degenerates into three independent simple pendulums.

### 11.6.2 Masses concentrated along a cord

A cord is a one dimensional element of mechanics, which can only transmit a force of traction. We are going to deal with the example of lateral and coplanar vibrations of $n$ point masses on a cord with an initial tension $T$. The behaviour of the system is linear when the influence of the movements of the masses on the initial tension is negligible. Assuming the cord to be massless, the displacement consists of $n + 1$ straight segments (figure 11.4).

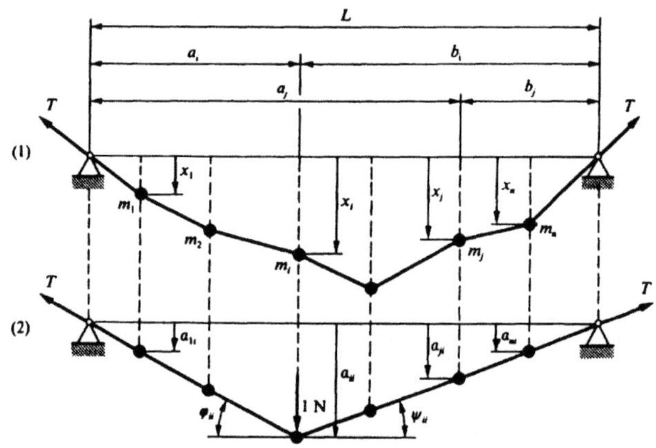


Fig. 11.4 Point masses on a cord with tension T
1) dynamic displacement
2) static displacement due to a load of 1 N on $m_i$

It is convenient to use the form (11.18) of the differential equation

$$[\alpha]\ [M]\ \ddot{\vec{x}} + \vec{x} = \vec{0}$$

The mass matrix is diagonal

$$[M] = \begin{bmatrix} m_1 & & 0 \\ & \ddots & \\ 0 & & m_n \end{bmatrix}$$

One calculates the matrix of influence coefficients by considering the equilibrium of the mass $m_i$ subject to a load of 1 N . It gives, the angles being small,

$$1 = T(\varphi_{ii} + \psi_{ii}) = T(\frac{\alpha_{ii}}{a_i} + \frac{\alpha_{ii}}{b_i})$$

With $a_i + b_i = L$ one obtains

$$\alpha_{ii} = \frac{a_i\ b_i}{T\ L}$$

Then, by simple proportionality

$$\begin{cases} \alpha_{ji} = \dfrac{b_j}{b_i}\,\alpha_{ii} & \text{if } j > i\ (a_j > a_i) \\ \alpha_{ji} = \dfrac{a_j}{a_i}\,\alpha_{ii} & \text{if } j < i\ (a_j < a_i) \end{cases}$$

whence by replacing $\alpha_{ii}$ by its value

$$\begin{cases} \alpha_{ji} = \dfrac{a_i\ b_j}{T\ L} & \text{if } j > i\ \ (m_j \text{ to the right of } m_i) \\ \alpha_{ji} = \dfrac{a_j\ b_i}{T\ L} & \text{if } j < i\ \ (m_j \text{ to the left of } m_i) \end{cases} \qquad (11.101)$$

One notes that the influence coefficients, and therefore the natural frequencies of small transverse movements of the system, are not a function of the cross-section nor of the modulus of elasticity of the cord. Apart from the lengths and the masses, only the initial tension is important.

The situation would be reversed if one studied the small longitudinal movements of the system.

Let us examine the particular case of three equidistant equal masses. The mass matrix is then (figure 11.5)

$$[M] = m\,[I]$$

One knows that the matrix of influence coefficients $[\alpha]$ is symmetrical because of the linearity of the system. As in the previous example of the triple pendulum, the system has an axis of geometrical symmetry and the matrix thus becomes doubly symmetric.

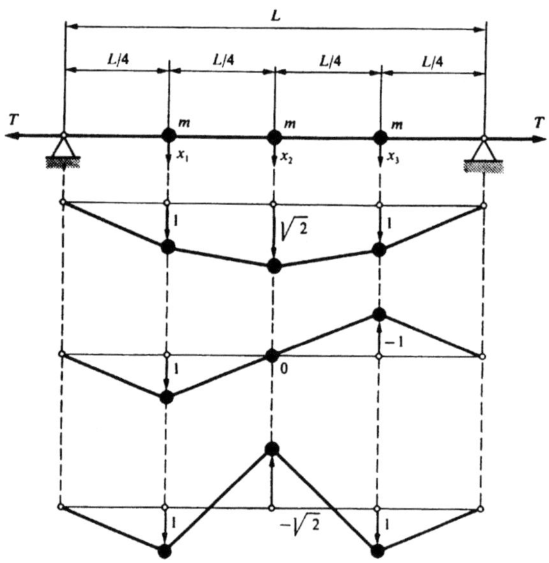


Fig. 11.5 System of three equal masses on a cord of tension T

One easily finds by means of relations (11.101)

$$[\alpha] = \frac{L}{4\,T} \begin{bmatrix} 3/4 & 1/2 & 1/4 \\ 1/2 & 1 & 1/2 \\ 1/4 & 1/2 & 3/4 \end{bmatrix} \tag{11.102}$$

The matrix $[E] = [\alpha]\,[M]$ (inverse of the core) is calculated immediately in this specific case and one chooses the form (11.22) of the characteristic equation ($\tau = 1/\delta$ is the frequency function)

$$\left| [E] - \tau [I] \right| = 0$$

$$\left| \frac{m\,L}{4\,T} \begin{bmatrix} 3/4 & 1/2 & 1/4 \\ 1/2 & 1 & 1/2 \\ 1/4 & 1/2 & 3/4 \end{bmatrix} - \tau \begin{bmatrix} 1 & 0 & 0 \\ 0 & 1 & 0 \\ 0 & 0 & 1 \end{bmatrix} \right| = 0$$

With the more convenient notation

$$z = \tau \frac{4\,T}{m\,L} = \frac{1}{\delta} \frac{4\,T}{m\,L} = \frac{1}{\omega^2} \frac{4\,T}{m\,L} \Rightarrow \omega^2 = \frac{4}{z} \frac{T}{m\,L} \tag{11.103}$$

the equation becomes

$$\begin{vmatrix} (3/4 - z) & 1/2 & 1/4 \\ 1/2 & (1 - z) & 1/2 \\ 1/4 & 1/2 & (3/4 - z) \end{vmatrix} = 0 \tag{11.104}$$

One readily finds the three solutions and the corresponding natural angular frequencies

$$\begin{cases} z_1 = 1 + \dfrac{\sqrt{2}}{2} & \omega_1 = 1.53 \sqrt{\dfrac{T}{m\,L}} \\ z_2 = \dfrac{1}{2} & \omega_2 = 2.83 \sqrt{\dfrac{T}{m\,L}} \\ z_3 = 1 - \dfrac{\sqrt{2}}{2} & \omega_3 = 3.70 \sqrt{\dfrac{T}{m\,L}} \end{cases} \tag{11.105}$$

We will not reproduce the calculation of the natural mode shapes because it is trivial. The results are shown in figure 11.5. After having verified their orthogonality, one obtains the free state of the system

$$\begin{Bmatrix} x_1 \\ x_2 \\ x_3 \end{Bmatrix} = \begin{Bmatrix} 1 \\ \sqrt{2} \\ 1 \end{Bmatrix} X_1 \cos(\omega_1 t-\varphi_1) + \begin{Bmatrix} 1 \\ 0 \\ -1 \end{Bmatrix} X_2 \cos(\omega_2 t-\varphi_2) + \begin{Bmatrix} 1 \\ -\sqrt{2} \\ 1 \end{Bmatrix} X_3 \cos(\omega_3 t-\varphi_3) \qquad (11.106)$$

One notes that the central mass does not move in the second mode. This circumstance plays an important role in the steady state of the system [3].

### 11.6.3 Masses concentrated along a beam

For practical purposes, numerous real systems can be represented schematically, to a good approximation, by masses concentrated along a massless beam. They are then likened to discrete systems having a finite number of degrees of freedom, and thus a finite number of frequencies and natural modes.

We will limit ourselves here to the simplest case, that of the lateral and coplanar vibrations of point masses on a straight beam (figure 11.6)

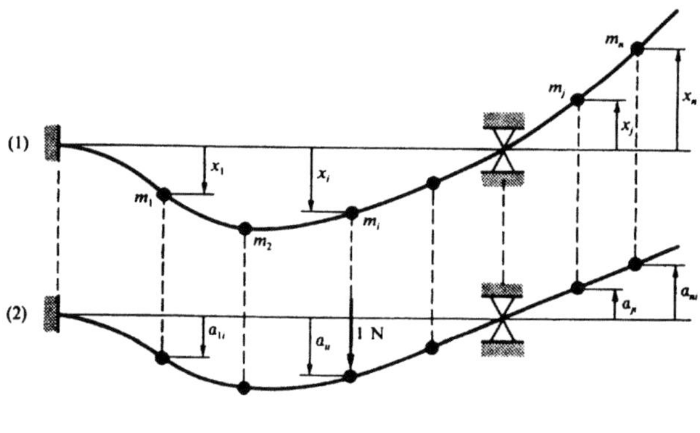


Fig. 11.6 Point masses along a beam
(1) dynamic displacement
(2) static displacement due to a load of 1 N on $m_i$

As in the preceding section, one uses equation (11.18)

$$[\alpha]\ [M]\ \ddot{\vec{x}} + \vec{x} = \vec{0}$$

In this particular case, the matrix [M] is known immediately whereas the matrix [α] must be established by means of the methods of the strength of materials.

Figure 11.7 shows the simplest case of a beam on two brackets supporting four equidistant equal masses. Since this system has a geometrical axis of symmetry, the flexibility matrix is doubly symmetrical.

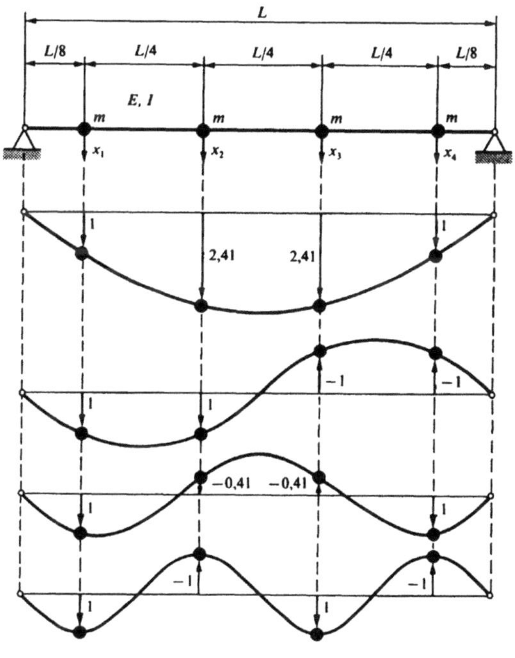


Fig. 11.7 System of four equal masses along a beam, fundamental angular frequency $\omega_1 = \frac{4.93}{L}\sqrt{\frac{E\ I}{m\ L}}$

The mass matrix is here

$$[M] = m\,[I]$$

A formula obtained from the theory of strength of materials allows one to calculate the matrix $[\alpha]$

$$[\alpha] = \frac{L^3}{122.88\ E\ I}\begin{bmatrix} 0.49 & 0.95 & 0.81 & 0.31 \\ 0.95 & 2.25 & 2.07 & 0.81 \\ 0.81 & 2.07 & 2.25 & 0.95 \\ 0.31 & 0.81 & 0.95 & 0.49 \end{bmatrix} \tag{11.107}$$

Let us return to the characteristic equation (11.22), $\tau = 1/\delta$ , being the frequency function,

$$\left|[E] - \tau[I]\right| = 0 \qquad [E] = [\alpha]\ [M]$$

In order to simplify the writing, one introduces the notation

$$z = \tau\,\frac{122.88\ E\ I}{m\ L^3} = \frac{1}{\delta}\,\frac{122.88\ E\ I}{m\ L^3} = \frac{1}{\omega^2}\,\frac{122.88\ E\ I}{m\ L^3} \tag{11.108}$$

whence one finds

$$\omega^2 = \frac{122.88}{z}\,\frac{E\ I}{m\ L^3} \quad \Rightarrow \quad \omega = \sqrt{\frac{122.88}{z}}\cdot\frac{1}{L}\cdot\sqrt{\frac{E\ I}{m\ L}} \tag{11.109}$$

The characteristic equation becomes thus

$$\begin{vmatrix} (0.49 - z) & 0.95 & 0.81 & 0.31 \\ 0.95 & (2.25 - z) & 2.07 & 0.81 \\ 0.81 & 2.07 & (2.25 - z) & 0.95 \\ 0.31 & 0.81 & 0.95 & (0.49 - z) \end{vmatrix} = 0 \tag{11.110}$$

The calculation of the solutions of this equation gives

$$\begin{cases} z_1 = 5.04900 & \omega_1 = \dfrac{4.9333}{L}\sqrt{\dfrac{E\ I}{m\ L}} \\ z_2 = 0.32000 & \omega_2 = 3.97\ \omega_1 \\ z_3 = 0.07098 & \omega_3 = 8.43\ \omega_1 \\ z_4 = 0.04000 & \omega_4 = 11.2\ \omega_1 \end{cases} \tag{11.111}$$

One then calculates the natural mode shapes, whose orthogonality is well established. They are shown plotted in figure 11.7. The free state is then determined

$$\begin{Bmatrix} x_1 \\ x_2 \\ x_3 \\ x_4 \end{Bmatrix} = \begin{Bmatrix} 1 \\ 2.41 \\ 2.41 \\ 1 \end{Bmatrix} X_1 \cos(\omega_1 t - \varphi_1) + \begin{Bmatrix} 1 \\ 1 \\ -1 \\ -1 \end{Bmatrix} X_2 \cos(\omega_2 t - \varphi_2) + \begin{Bmatrix} 1 \\ -0.41 \\ -0.41 \\ 1 \end{Bmatrix} X_3 \cos(\omega_3 t - \varphi_3) + \begin{Bmatrix} 1 \\ -1 \\ 1 \\ -1 \end{Bmatrix} X_4 \cos(\omega_4 t - \varphi_4) \quad (11.112)$$

It is interesting to compare the natural angular frequencies found here to those of a continuous beam with the same total mass 4 m. One shows that the natural angular frequencies of such a beam are given by the formula

$$\omega'_n = \frac{n^2 \pi^2}{L^2} \sqrt{\frac{E I}{\mu_1}} \quad (11.113)$$

in which $\mu_1$ is the mass per unit length and $n$ the order of the modes. The series of natural angular frequencies is thus

$$\omega'_1 = \frac{\pi^2}{L^2} \sqrt{\frac{E I}{\mu_1}} \qquad \omega'_2 = 4\, \omega'_1 \qquad \omega'_3 = 9\, \omega'_1 \qquad \omega'_4 = 16\, \omega'_1 \qquad \dots \quad (11.114)$$

If one considered the system of 4 masses of figure 11.7 as an approximation to a continuous beam one would have

$$\mu_1 = \frac{4\, m}{L} \quad \Rightarrow \quad m = \frac{\mu_1 L}{4} \quad \Rightarrow \quad \omega_1 = \frac{9.867}{L^2} \sqrt{\frac{E I}{\mu_1}} \quad (11.115)$$

By comparing (11.115) and (11.114), one sees that in this way one would make the following very small relative error for the fundamental angular frequency

$$\text{1st mode} \qquad \varepsilon_1 = \frac{\omega_1 - \omega'_1}{\omega'_1} = -\, 0.03\ \%$$

In the same way, the relative error for the angular frequencies of the other modes would be

| | | |
|---|---|---|
| 2nd | mode | $\varepsilon_2 = -0.73\ \%$ |
| 3rd | mode | $\varepsilon_3 = -6.32\ \%$ |
| 4th | mode | $\varepsilon_4 = -29.8\ \%$ |

The relative error increases rapidly with the order of the modes, it is about - 30 % for the fourth. This is easy to understand, because the dynamic displacements of the discrete system deviate increasingly from those of the continuous beam as the number of inertial forces (one per point mass) becomes smaller on each undulation. This number is four for the first natural mode shape and one for the fourth.

The natural mode shapes of a uniform continuous beam on two supports are pure sine waves. One notes that the first natural mode shape of the system of four masses is almost exactly a semi-sine wave.

Rayleigh quotient

In the particular case examined, the fundamental eigenvector is given by the expression

$$\vec{\beta}_1^T = \{1 \quad 2.41 \quad 2.41 \quad 1\} \quad \Rightarrow \quad \frac{\beta_{21}}{\beta_{11}} = \frac{2.41}{1} = 2.41 \tag{11.116}$$

Let us suppose that one has chosen, in order to calculate the Rayleigh coefficient, the following approximate vector

$$\vec{u}^T = \{\ 1 \quad 2 \quad 2 \quad 1\ \} \quad \Rightarrow \quad \frac{u_2}{u_1} = \frac{2}{1} = 2.00 \tag{11.117}$$

This is a large enough approximation; in effect, the relative error made can be estimated as follows

$$\varepsilon' = \frac{2.00 - 2.41}{2.41} = -17\ \%$$

Let us calculate an approximate value $\omega_a$ of the fundamental angular frequency by means of the relation (11.89)

$$\omega_a^2 = R(u) = \frac{\vec{u}^T \ \vec{u}}{\vec{u}^T[\alpha] \ [M] \ \vec{u}} = \frac{N}{D}$$

One determines in the first place the value of the numerator using (11.117)

$$N = \{ 1 \quad 2 \quad 2 \quad 1 \} \begin{Bmatrix} 1 \\ 2 \\ 2 \\ 1 \end{Bmatrix} = 10$$

then that of the denominator by using the matrices $[M] = m\,[I]$ and $[\alpha]$ given by (11.107)

$$D = \{ 1 \quad 2 \quad 2 \quad 1 \} \, [\alpha] \, m \, [I] \begin{Bmatrix} 1 \\ 2 \\ 2 \\ 1 \end{Bmatrix} = \frac{m\,L^3}{122.88\ E\ I} \cdot 50.24$$

One obtains thus

$$\omega_a^2 = 122.88 \, \frac{E\ I}{m\ L^3} \, \frac{10}{50.24} = 24.46 \, \frac{E\ I}{m\ L^3}$$

$$\omega_a = 4.946 \, \frac{1}{L} \sqrt{\frac{E\ I}{m\ L}}$$

The relative error made on the fundamental eigenvalue is then

$$\varepsilon'' = \frac{\omega_a^2 - \omega_1^2}{\omega_1^2} = \frac{24.46 - 24.33}{24.33} = 0.53\ \%$$

In accordance with the theory, the above results show that the approximate natural angular frequency is larger than the exact eigenvalue, and that if $\varepsilon'$ is of the first order, $\varepsilon''$ is of the second order, $\varepsilon'' = 0.031\,|\varepsilon'|$ .

Finally, on the natural angular frequency itself, the relative error is again smaller

$$\varepsilon'' = \frac{\omega_a - \omega_1}{\omega_1} = \frac{4.946 - 4.933}{4.933} = 0.26\ \%$$

Clearly one could have chosen a better approximation for $u$ than (11.117). Two possibilities immediately come to mind :

- the first natural mode shape is a sine wave which gives, with the normalization $u_1 = 1$

$$\vec{u}^T = \{\, 1 \quad 2.414 \quad 2.414 \quad 1 \,\} \tag{11.118}$$

- the first natural mode shape is the static displacement after normalization, in other words

$$u_i = \sum_j^4 \alpha_{ij} \Big/ \sum_j^4 \alpha_{1j}$$

hence

$$\vec{u}^T = \{\, 1 \quad 2.375 \quad 2.375 \quad 1 \,\} \tag{11.119}$$

In these two cases, one confirms that the relative error on the natural angular frequency becomes completely negligible ($\varepsilon'' < 0.01\ \%$).

### 11.6.4 Study of the behaviour of a milling table

Figure 11.8 shows schematically all the moving parts of a small scale milling table.

The actual table, of mass $m_t$ , is driven by a direct current motor of inertia $J_m$ , by means of a ribbed belt and a ballscrew of pitch $t$ . The inertia of each of the identical pulleys which support the belt, is designated by $J_p$, and the intrinsic inertia of the screw of mass $m_v$ , by $J_v$ . Also let $k_c$ be the traction stiffness of the belt, $k_b$ and $k_e$ the stiffnesses of the stop and of the screw nut respectively. The screw itself is subject to tension-compression and torsional forces which we assume to be decoupled. Let us designate the corresponding stiffnesses by $k_{vc}$ and $k_{vt}$ .

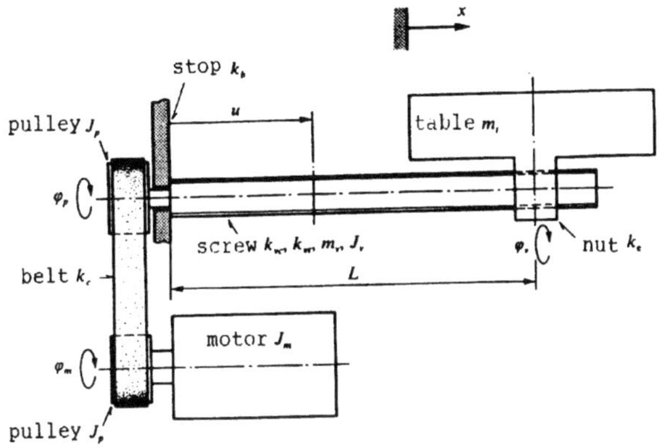


Fig. 11.8 Diagram of all the massive moving parts of a milling table

The equations of the system can be established using the following generalized coordinates,

- x  linear displacement of the table
- $\varphi_v$  rotation of the screw to the right of the nut
- $\varphi_p$  rotation of the pulley on the end of the screw
- $\varphi_m$  rotation of the motor pulley

This choice implies that the motor shaft as well as the pulleys are considered undeformable. The origin of the generalized coordinates corresponds to the rest position of the system.

By also defining the active length $L$ of the screw, the linear displacement per radian $\alpha = t/2\pi$ of the nut, and the radius $R_p$ of the pulleys, we can calculate the potential and kinetic energies of the system in order to use Lagrange's equations.

Kinetic energy

· Table

$$T_t = \frac{1}{2} m_t \dot{x}^2$$

· Screw, translation of

For the screw, one assumes that the longitudinal and angular velocities vary linearly from one extremity to the other. This gives, u being the auxiliary variable of integration indicated in the figure,

$$T_{vt} = \frac{1}{2} \frac{m_v}{L} \int_0^L \left[\frac{u}{L} (\dot{x} - \alpha \dot{\varphi}_v)\right]^2 du$$

$$T_{vt} = \frac{1}{2} \frac{m_v}{3} (\dot{x} - \alpha \dot{\varphi}_v)^2$$

· Screw, rotation of

$$T_{vr} = \frac{1}{2} \frac{J_v}{L} \int_0^L \left[\dot{\varphi}_p + \frac{u}{L} (\dot{\varphi}_v - \dot{\varphi}_p)\right]^2 du$$

$$T_{vr} = \frac{1}{2} \frac{J_v}{3} (\dot{\varphi}_p^2 + \dot{\varphi}_v^2 + \dot{\varphi}_p \dot{\varphi}_r)$$

· Pulley at end of screw

$$T_p = \frac{1}{2} J_p \dot{\varphi}_p^2$$

· Motor and associated pulley

$$T_m = \frac{1}{2} (J_m + J_p) \dot{\varphi}_m^2$$

Neglecting the kinetic energy in the belt, the total kinetic energy is then

$$T = T_t + T_{vt} + T_{vr} + T_p + T_m$$

This gives, by adding the above results,

$$T = \frac{1}{2}\,[m_t\,\dot{x}^2 + \frac{m_v}{3}\,(\dot{x} - \alpha\,\dot{\varphi}_v)^2 + \frac{J_v}{3}\,(\dot{\varphi}_p^2 + \dot{\varphi}_v^2 + \dot{\varphi}_p\,\dot{\varphi}_v) + J_p\,\dot{\varphi}_p^2 + (J_m + J_p)\,\dot{\varphi}_m^2] \qquad (11.120)$$

Potential energy

· Tension-compression of the screw, taking account of the stiffnesses of the stop and of the nut

$$V_{vc} = \frac{1}{2}\,k\,(x - \alpha\,\varphi_v)^2$$

In this expression, $k$ is the equivalent stiffness defined as follows

$$k = (\frac{1}{k_{vc}} + \frac{1}{k_b} + \frac{1}{k_e})^{-1} \qquad (11.121)$$

· Torsion of the screw

$$V_{vt} = \frac{1}{2}\,k_{vt}(\varphi_v - \varphi_p)^2$$

· Tension in the belt

$$V_c = \frac{1}{2}\,k_c\,R_p^2(\varphi_p - \varphi_m)^2$$

The total potential energy is the sum

$$V = V_{vc} + V_{vt} + V_c$$

which has the value, using the above expressions,

$$V = \frac{1}{2}\,[k(x - \alpha\,\varphi_v)^2 + k_{vt}(\varphi_v - \varphi_p)^2 + k_c\,R_p^2(\varphi_p - \varphi_m)^2] \qquad (11.122)$$

After differentiating equations (11.120) and (11.122) using Lagrange's equations, the differential equations in matrix form can be written

$$\begin{bmatrix} m_t + m_v/3 & -\alpha\, m_v/3 & 0 & 0 \\ -\alpha\, m_v/3 & \alpha^2\, m_v/3 + J_v/3 & J_v/6 & 0 \\ 0 & J_v/6 & J_v/3 + J_p & 0 \\ 0 & 0 & 0 & J_p + J_m \end{bmatrix} \begin{Bmatrix} \ddot{x} \\ \ddot{\varphi}_r \\ \ddot{\varphi}_p \\ \ddot{\varphi}_m \end{Bmatrix} +$$

$$\begin{bmatrix} k & -\alpha\, k & 0 & 0 \\ -\alpha\, k & \alpha^2\, k + k_{vt} & -k_{vt} & 0 \\ 0 & -k_{vt} & k_{vt} + k_c\, R_p^2 & -k_c\, R_p^2 \\ 0 & 0 & -k_c\, R_p^2 & k_c\, R_p^2 \end{bmatrix} \begin{Bmatrix} x \\ \varphi_v \\ \varphi_p \\ \varphi_m \end{Bmatrix} = \begin{Bmatrix} 0 \\ 0 \\ 0 \\ 0 \end{Bmatrix} \qquad (11.123)$$

The differential system being established, it is necessary to introduce the numerical values for the elements of the milling table studied, in order to continue the calculation.

<u>Numerical values</u>

· Ballscrew

$m_v = 4$ kg $\qquad t = 5\cdot10^{-3}$ m

$J_v = 4.34\cdot10^{-4}$ kgm$^2$ $\qquad d = 2.9\cdot10^{-2}$ m

$L = 0.7$ m $\qquad E = 2.1\cdot10^{11}$ N/m$^2$

$G = 0.8\cdot10^{11}$ N/m$^2$

One can thus calculate

$$\alpha = t/2\pi = 7.96\cdot10^{-4} \text{ m/rad}$$

$$k_{vc} = \frac{E}{L}\,\frac{\pi d^2}{4} = 1.98\cdot10^{8} \text{ N/m}$$

$$k_{vt} = \frac{G}{L}\,\frac{\pi d^4}{32} = 7.94\cdot10^{3} \text{ Nm/rad}$$

· Stop $\qquad k_b = 2.8\cdot10^{9}$ N/m (measured)

· Nut $\quad k_e = 9.2 \cdot 10^8$ N/m (measured)

· Equivalent stiffness from (11.121) $\quad k = 1.54 \cdot 10^8$ N/m

· Belt $\quad k_c = 10^6$ N/m (measured)

· Pulley $\quad R_p = 4.3 \cdot 10^{-2}$ m $\qquad k_c\, R_p^2 = 1.85 \cdot 10^3$ Nm/rad
$\qquad J_p = 1.45 \cdot 10^{-3}$ kgm²

· Mass of the table $\quad m_t = 170$ kg

· Inertia of the motor $\quad J_m = 1.4 \cdot 10^{-3}$ kgm²

With the above values, the matrices of the system become

$$[M] = \begin{bmatrix} 1.713 \cdot 10^2 & -1.061 \cdot 10^{-3} & 0 & 0 \\ -1.061 \cdot 10^{-3} & 1.455 \cdot 10^{-4} & 7.233 \cdot 10^{-5} & 0 \\ 0 & 7.233 \cdot 10^{-5} & 1.595 \cdot 10^{-3} & 0 \\ 0 & 0 & 0 & 2.850 \cdot 10^{-3} \end{bmatrix} \qquad (11.124)$$

$$[K] = \begin{bmatrix} 1.540 \cdot 10^8 & -1.225 \cdot 10^5 & 0 & 0 \\ -1.225 \cdot 10^5 & 8.038 \cdot 10^3 & -7.940 \cdot 10^3 & 0 \\ 0 & -7.940 \cdot 10^3 & 9.789 \cdot 10^3 & -1.849 \cdot 10^3 \\ 0 & 0 & -1.849 \cdot 10^3 & 1.849 \cdot 10^3 \end{bmatrix} \qquad (11.125)$$

One then calculates the eigenvalues, angular frequencies and natural frequencies

$$\begin{cases} \delta_1 = 0 & \omega_1 = 0 & f_1 = 0 \\ \delta_2 = 8.722 \cdot 10^5 \text{ s}^{-2} & \omega_2 = 9.339 \cdot 10^2 \text{ s}^{-1} & f_2 = 148 \text{ Hz} \\ \delta_3 = 1.696 \cdot 10^6 \text{ s}^{-2} & \omega_3 = 1.302 \cdot 10^3 \text{ s}^{-1} & f_3 = 207 \text{ Hz} \\ \delta_4 = 6.683 \cdot 10^7 \text{ s}^{-2} & \omega_4 = 8.175 \cdot 10^3 \text{ s}^{-1} & f_4 = 1301 \text{ Hz} \end{cases} \qquad (11.126)$$

These results show that there exists a mode with zero frequency, in other words a **zero mode** (or **rigid body** mode) which we are going to return to later.

The change of basis matrix $[B]$, which consists of the modal vectors $\vec{\beta}_p$ normalized in such a way that $[B]^T[M]\,[B] = [I]$, is then written

$$[B] = \begin{bmatrix} 1.144\cdot10^{-2} & 7.379\cdot10^{-2} & 1.614\cdot10^{-2} & 3.680\cdot10^{-4} \\ 1.437\cdot10^{1} & 2.754 & -\ 1.825\cdot10^{1} & -\ 8.052\cdot10^{1} \\ 1.437\cdot10^{1} & 1.601 & -\ 1.787\cdot10^{1} & 1.063\cdot10^{1} \\ 1.437\cdot10^{1} & -\ 4.649 & 1.108\cdot10^{1} & -\ 1.042\cdot10^{-1} \end{bmatrix} \qquad (11.127)$$

Given the non-homogeneity of the initial generalized coordinates, that is to say a linear displacement and three rotations, the ratios $\beta_{ip}/\beta_{1p}$ are not dimensionless. In effect

$$[\beta_{ip}/\beta_{1p}] = \text{rad/m}$$

Let us note also that the modal vector corresponding to the zero mode introduces the ratios

$$\beta_{ip}/\beta_{1p} = \frac{1}{\alpha}$$

This confirms the fact that this mode represents a simple sliding of the system, without putting into play any energy of deformation.

It is also necessary to note that in the example considered, the lowest non-zero natural frequency, that is to say $f_2$, is determined almost uniquely by the relatively weak stiffness $k$, of the stop - ballscrew - nut assembly. In effect, one obtains a value very close to $f_2$ by considering this stiffness and the mass of the table as an elementary oscillator

$$f_2' = \frac{1}{2\pi}\sqrt{\frac{k}{m_t}} = 151 \text{ Hz}$$

This result shows that in order to improve the vibratory behaviour of a milling table, in particular to increase the lowest of the natural frequencies, it is most important to seek as large as possible a stiffness for the stop - ballscrew - nut assembly. The possibilities for reducing the mass of the table are more limited.

Suppression of the zero mode

When one knows a priori that a system is semi-definite, that is to say when one knows the sliding eigenvector, it is always possible to suppress one of the generalized coordinates, which makes the zero natural angular frequency disappear. One obtains in this way a new system, whose order is reduced by one, but which is definite.

Let us assume that in the preceding example we have noted that

$$[K]\begin{Bmatrix}\alpha\\1\\1\\1\end{Bmatrix}=\vec{0} \qquad (11.128)$$

We should have then been able to say that this vector was the eigenvector $\vec{\beta}_0$ of a mode with zero angular frequency. By using the orthogonality relationship for the natural modes, it gives

$$\vec{\beta}_0^T[M]\,\vec{\beta}_r=\vec{0} \qquad (11.129)$$

Not knowing $\vec{\beta}_r$ yet, one can write

$$\vec{\beta}_r^T=\{x\ \varphi_v\ \varphi_p\ \varphi_m\} \qquad (11.130)$$

and the orthogonality above leads to the equation

$$\alpha\, m_t\, x + J_v/2\ \varphi_v + (J_p + J_v/2)\ \varphi_p + (J_p + J_m)\ \varphi_m = 0 \qquad (11.131)$$

This relation, differentiated with respect to time, expresses the conservation of momentum of the system. Furthermore, it enables one to extract one or other of the initial coordinates as a function of the remaining coordinates, for example $\varphi_V$

$$\varphi_V = - \frac{2 \alpha m_t}{J_V} x - (1 + \frac{2 J_p}{J_V}) \varphi_p - \frac{2(J_p + J_m)}{J_V} \varphi_m \tag{11.132}$$

One can in this way reduce the order of the system of equations by means of the transformation

$$\begin{Bmatrix} x \\ \varphi_V \\ \varphi_p \\ \varphi_m \end{Bmatrix} = \begin{bmatrix} 1 & 0 & 0 \\ - \frac{2 \alpha m_t}{J_V} & - (1 + \frac{2 J_p}{J_V}) & - \frac{2(J_p + J_m)}{J_V} \\ 0 & 1 & 0 \\ 0 & 0 & 1 \end{bmatrix} \begin{Bmatrix} x \\ \varphi_p \\ \varphi_m \end{Bmatrix} \tag{11.133}$$

The remaining system is then of order $n = 3$ . By solving it one obtains the three modes with non-zero angular frequencies.

# CHAPTER 12 FREE STATE OF THE GENERALIZED DISSIPATIVE OSCILLATOR

The system of differential equations for the dissipative free state, in matrix form, is that of relation (10.1) with the right-hand side set to zero.

$$[M]\,\ddot{\vec{x}} + [C]\,\dot{\vec{x}} + [K]\,\vec{x} = \vec{0} \tag{12.1}$$

The existence of the dissipative terms, shown by the loss matrix [C] , considerably complicates the problem. We must adopt different methods to solve this equation, depending on the nature of this matrix. As in the previous chapter, the solutions will introduce the concept of vibratory modes of the oscillating system.

## 12.1 Limits of classical modal analysis

By adopting the same steps as section (11.3), we determine the new set of coordinates q(t) , from the set x(t) by the relation

$$\vec{x} = [B]\,\vec{q} \tag{12.2}$$

and which allows one to decouple the differential system (12.1).

Let us therefore make this change of variables. It gives

$$[M]\,[B]\,\ddot{\vec{q}} + [C]\,[B]\,\dot{\vec{q}} + [K]\,[B]\,\vec{q} = \vec{0} \tag{12.3}$$

Let us then pre-multiply by $[B]^T$ in such a way as to restore the symmetry of the matrices, whence

$$[B]^T[M]\,[B]\,\ddot{\vec{q}} + [B]^T[C]\,[B]\,\dot{\vec{q}} + [B]^T[K]\,[B]\,\vec{q} = \vec{0} \tag{12.4}$$

In order to simplify the writing, one adopts the notation

$$\begin{cases} [M'] = [B]^T[M]\,[B] \\ [C'] = [B]^T[C]\,[B] \\ [K'] = [B]^T[K]\,[B] \end{cases} \tag{12.5}$$

Equation (12.4) becomes

$$[M']\,\ddot{\vec{q}} + [C']\,\dot{\vec{q}} + [K']\,\vec{q} = \vec{0} \tag{12.6}$$

So, as a result of the damping, the problem is to determine if there exists a matrix [B] allowing one to diagonalize simultaneously not just two, but three symmetric matrices.

If that is the case, then the initial system (12.1) can be replaced by a system of $n$ equations analogous to that of a dissipative elementary oscillator, that is to say

$$m_p^\circ\,\ddot{q}_p + c_p^\circ\,\dot{q}_p + k_p^\circ\,q_p = 0 \qquad p = 1, 2, \ldots, n \tag{12.7}$$

Let us return to equation (12.6) and pre-multiply it by $[M']^{-1}$

$$\ddot{\vec{q}} + [M']^{-1}[C']\,\dot{\vec{q}} + [M']^{-1}[K']\,\vec{q} = \vec{0} \tag{12.8}$$

If the possibility exists to simultaneously diagonalize [M] , [C] and [K] , then the matrices $[M']^{-1}[C']$ and $[M']^{-1}[K']$ of (12.8) are diagonal and their product is commutative

$$[M']^{-1}[C']\,[M']^{-1}[K'] = [M']^{-1}[K']\,[M']^{-1}[C'] \tag{12.9}$$

Taking account of the relations (12.5), the equality above takes the form

$$[B]^{-1}[M]^{-1}[C]\,[M]^{-1}[K]\,[B] = [B]^{-1}[M]^{-1}[K]\,[M]^{-1}[C]\,[B] \tag{12.10}$$

By pre-multiplying the two sides by [M] [B] and by post-multiplying them by $[B]^{-1}$ , one finally obtains

$$[C]\,[M]^{-1}[K] = [K]\,[M]^{-1}[C] \tag{12.11}$$

This relation, known as the **Caughey condition** [1], is necessary and sufficient in order that the initial system (12.1) can be decoupled, in other words be reduced to $n$ elementary equations (12.7), by means of a change of basis defined by the constant and real matrix [B] .

When the Caughey condition is satisfied, the algorithm of section 11.3 is applicable and the solutions of the differential system (12.1) consist of $n$ damped **vibratory modes,** described as **classic** or **real** , as we will establish in the following section. In the opposite case, the change of basis must be made in another space, called **phase-space,** the matrix [B] is complex and of dimension $2n$ . The formulation of the problem is then established by means of **hamiltonian mechanics** or the **transformation of Duncan** (section 12.5).

One can also show [1] that condition (12.11) is satisfied if the damping matrix meets to the following condition, which is sufficient but not necessary :

$$[M']^{-1}[C'] = \sum_{i=0}^{n-1} \alpha_i \, [[M']^{-1}[K\,]]^i \qquad (12.12)$$

In this relation, $n$ is the order of the matrices [M], [C], [K] and the $\alpha_i$ are any real coefficients. As an example, let us consider the case where only $\alpha_0$ and $\alpha_1$ are different from zero.

$$[M']^{-1}[C'] = \alpha_0 \, [I] + \alpha_1 \, [M']^{-1}[K']$$

Let us pre-multiply by [M'] and go back to the initial matrices by means of the relations (12.5), then (12.13)

$$[C] = \alpha_0 \, [M] + \alpha_1 \, [K]$$

One obtains in this way a dissipation matrix corresponding to the **proportional damping.**

## 12.2 Dissipative free state with real modes

Let us assume that the Caughey condition (12.11) is satisfied. Then the change of basis matrix [B], which is determined as in section 11.3, diagonalizes [M] , [C] and [K] simultaneously.

The differential system (12.4) takes the form

$$[M°]\ \ddot{\vec{q}} + [C°]\ \dot{\vec{q}} + [K°]\ \vec{q} = \vec{0}$$

It consists of 3 diagonal matrices defined as follows

$$\begin{aligned} [M°] &= [B]^T[M]\ [B] \\ [C°] &= [B]^T[C]\ [B] \\ [K°] &= [B]^T[K]\ [B] \end{aligned} \tag{12.14}$$

This system is composed of n independent equations analogous to that of a dissipative linear elementary oscillator (12.7) (12.15)

$$m_p^\circ\ \ddot{q}_p + c_p^\circ\ \dot{q}_p + k_p^\circ\ q_p = 0 \qquad p = 1, 2, \ldots, n$$

then, by dividing by the mass,

$$\ddot{q}_p + \frac{c_p^\circ}{m_p^\circ}\ \dot{q}_p + \frac{k_p^\circ}{m_p^\circ}\ q_p = 0 \qquad p = 1, 2, \ldots, n$$

(12.16)

Let us define the quantities

$$\begin{cases} \omega_{op}^2 = \dfrac{k_p^\circ}{m_p^\circ} \\ 2\lambda_p = \dfrac{c_p^\circ}{m_p^\circ} \end{cases} \tag{12.17}$$

which are respectively the elements of the diagonal matrices

$$\begin{aligned} \lceil \Omega_0^2 \rfloor &= \lceil \Delta \rfloor = [B]^{-1}[M]^{-1}[K]\ [B] \\ \lceil 2\Lambda \rfloor &= [B]^{-1}[M]^{-1}[C]\ [B] \end{aligned} \tag{12.18}$$

This gives finally

$$\ddot{q}_p + 2\ \lambda_p\ \dot{q}_p + \omega_{0p}^2\ q_p = 0 \qquad p = 1,\ 2,\ \ldots,\ n \tag{12.19}$$

For such a system, the two matrices defined in (12.18) are determined on the basis of an eigenvalue problem in the sense of section 11.3, using the matrices [M] and [K] or the matrices [M] and [C] . By analogy with the elementary oscillator, the quantity $\lambda_p$ is called the **modal damping coefficient** and the ratio

$$\eta_p = \frac{\lambda_p}{\omega_{0p}} \tag{12.20}$$

the **modal damping factor** or **relative modal damping.**

The integration of equations (12.19) gives immediately

$$q_p = Q_p\ e^{-\lambda_p t} \cos(\omega_p t - \varphi_p) \qquad p = 1,\ 2,\ \ldots,\ n \tag{12.21}$$

$\omega_p$ being the **natural angular frequency with damping** which is defined as follows

$$\omega_p = \sqrt{\omega_{0p}^2 - \lambda_p^2} = \omega_{0p} \sqrt{1 - \eta_p^2} \tag{12.22}$$

Let us now return to the initial coordinates of the system by means of (12.2)

$$\vec{x} = [B]\,\vec{q}$$

One obtains as before

$$\vec{x} = \sum_{p}^{n} \vec{B}_p\, q_p$$

then, by using (12.21)

$$\vec{x} = \sum_{p}^{n} \vec{B}_p\, Q_p\, e^{-\lambda_p t} \cos(\omega_p t - \varphi_p) \qquad (12.23)$$

The components of the vectors $\vec{B}_p$ being unnormalized, one can adopt, as before, the change of notation $\vec{B}_p\, Q_p = \vec{\beta}_p\, X_p$, whence finally

$$\vec{x} = \sum_{p}^{n} \underbrace{\vec{\beta}_p\, X_p\, e^{-\lambda_p t} \cos(\omega_p t - \varphi_p)}_{\text{pth natural mode}} \qquad (12.24)$$

The eigenvectors $\vec{\beta}_p$ of the matrix [B] are identical to those of section 11.3, having the same orthogonality properties, and they can be normalized in the same way.

To summarize, when the Caughey condition is satisfied, the solutions of the dissipative system have the same structure as those of a conservative system apart from the damping. The damping factor is different for each natural mode.

## 12.3 Response to an initial excitation in the case of real modes

For a release with any initial conditions

$$\vec{x}(0) = \vec{X}_0 \tag{12.25}$$

$$\dot{\vec{x}}(0) = \vec{V}_0 \tag{12.26}$$

the solution (12.24) leads to the relations

$$\vec{X}_0 = \sum_p^n \vec{\beta}_p X_p \cos \varphi_p \tag{12.27}$$

$$\vec{V}_0 = \sum_p^n \vec{\beta}_p X_p(\omega_p \sin \varphi_p - \lambda_p \cos \varphi_p) \tag{12.28}$$

By pre-multiplying these equations by $\vec{\beta}_r^T[M]$ and by using the orthogonality relations (11.58), which are still valid here, we get

$$X_r \cos \varphi_r = \frac{1}{m_r^\circ} \vec{\beta}_r^T[M] \vec{X}_0 \tag{12.29}$$

$$X_r(\omega_r \sin \varphi_r - \lambda_r \cos \varphi_r) = \frac{1}{m_r^\circ} \vec{\beta}_r^T[M] \vec{V}_0$$

then, by isolating $X_r \sin \varphi_r$ in this last relation

$$X_r \sin \varphi_r = \frac{1}{m_r^\circ \omega_r} \vec{\beta}_r^T[M] \{\vec{V}_0 + \lambda_r \vec{X}_0\} \tag{12.30}$$

The response of the system can then be written

$$\vec{x} = \sum_p^n \frac{1}{m_p^\circ} \vec{\beta}_p e^{-\lambda_p t} (\vec{\beta}_p^T[M] \vec{X}_0 \cos \omega_p t + \\ + \frac{1}{\omega_p} \vec{\beta}_p^T[M] \{\vec{V}_0 + \lambda_p \vec{X}_0\} \sin \omega_p t) \tag{12.31}$$

Finally, one easily confirms that the initial conditions which isolate a mode of vibration are analogous to those put forward in section 11.3.

## 12.4 General case

Let us go back to equation (12.1)

$$[M]\,\ddot{\vec{x}} + [C]\,\dot{\vec{x}} + [K]\,\vec{x} = \vec{0}$$

and let us search for a solution of the type

$$\vec{x} = \vec{X}\,e^{-\delta t}$$

where, in the sense of section 11.3

$$\vec{x} = \vec{B}\,e^{-\delta t} \tag{12.32}$$

By introducing this solution into the differential equation, one obtains

$$[\delta^2[M] - \delta\,[C] + [K]]\,\vec{B} = \vec{0}$$

In order to simplify the notation, let us write

$$[N] = [\delta^2[M] - \delta\,[C] + [K]] \tag{12.33}$$

Which gives thus

$$[N]\,\vec{B} = \vec{0} \tag{12.34}$$

The matrix $[N]$, which is a function of $\delta$, is square and symmetric. The homogeneous system (12.34) only has non-trivial solutions if its determinant is zero, that is to say

$$\left|\delta^2[M] - \delta\,[C] + [K]\right| = 0 \tag{12.35}$$

This relation, which is called the **characteristic equation** like similar relations encountered previously, is of order $2n$ in $\delta$. Its roots can be real, complex or purely imaginary.

- The **real roots** are necessarily positive and correspond to some decreasing aperiodic displacements. If all the roots are of this type, the system would no longer belong to vibratory mechanics.

- The **complex roots** are conjugated in pairs, with a positive real part. The modal vectors which are attached to them are also complex conjugates and correspond, after combining, to the damped oscillatory modes.

- The purely **imaginary roots,** conjugated in pairs, correspond to the conservative oscillatory modes. The existence of such modes, rare in practice, is only possible for certain very specific configurations of the damping matrix $[C]$.

We are going to assume in what follows that all the roots are complex. In effect, the purely imaginary roots are only a particular case, while the real roots hardly present any interest in modal analysis. Moreover, in order to simplify the writing, the complex quantities will no longer be underlined, contrary to the convention adopted for chapters 5 and 6. In effect, at this stage of the presentation the reader will recognize without difficulty the nature of the quantities used.

By assuming that all the eigenvalues are distinct, let us write a particular solution for the differential system.

$$\vec{x} = \vec{B}_p \, e^{-\delta_p t} \tag{12.36}$$

The complex column-vector $\vec{B}_p$ is a modal vector of the system. It is equal to one of the non-zero column-vectors of the adjoint matrix of $[N]$, in which one has replaced $\delta$ by $\delta_p$.

The general solution for the system is obtained by linear combination of the solution (12.36).

$$\vec{x} = \sum_{p}^{2n} \gamma_p \vec{x}_p = \sum_{p}^{2n} \gamma_p \vec{B}_p e^{-\delta_p t} \qquad (12.37)$$

The arbitrary constants $\gamma_p$ are complex in the general case. In matrix form, one can write

$$\vec{x} = [B] \{\gamma_p e^{-\delta_p t}\} \qquad (12.38)$$

The matrix [B] cannot be used for a change of basis of the type

$$\vec{x} [B] \vec{q}$$

for it is rectangular, of order $n \times 2n$, which makes the converse transformation impossible (in particular, this prevents one from solving the problems of forced states).

The best way to surmount this difficulty is to convert the Lagrangian system of n second order differential equations into an equivalent system of 2n differential equations of the first order, called **Hamilton's canonical equations**. The n auxiliary variables are then the generalized momenta $p_i$ such that

$$\vec{p} = [M] \dot{\vec{x}} \qquad (12.39)$$

## 12.5 Hamiltonian equations for the system

The function of Lagrange - or Lagrangian - is the difference between the kinetic and potential energies

$$L = L(x, \dot{x}, t) - (T - V) \qquad (12.40)$$

One knows that the equations for the dynamics of the system are given by the following derivatives (equations of Lagrange)

$$\frac{d}{dt}\left(\frac{\partial L}{\partial \dot{x}_k}\right) - \frac{\partial L}{\partial x_k} = 0 \qquad k = 1, \ldots, n \tag{12.41}$$

The generalized momenta (12.39) are by definition

$$p_k = \frac{\partial L}{\partial \dot{x}_k} \qquad k = 1, \ldots, n \tag{12.42}$$

Let us recall that Hamiltonian mechanics, on the historical level, has been developed for conservative systems, as has Lagrangian mechanics also. We shall consequently establish Hamilton's equations for such systems, then we shall carry out the necessary modifications for the generalization of these equations to dissipative systems.

Let us use the **dual tranformation of Legendre** [15] so as to pass from a description of the dynamics of the system as a function of the variables $(x, \dot{x}, t)$ to a description as a function of the variables $(x, p, t)$.

One defines a new function for this transformation

$$H = \sum_{k}^{n} \frac{\partial L}{\partial \dot{x}_k} \dot{x}_k - L \tag{12.43}$$

which becomes, taking account of (12.42)

$$H = \sum_{k}^{n} p_k \dot{x}_k - L \tag{12.44}$$

It is then a matter of replacing the generalized velocities $\dot{x}_p$ by the corresponding momenta so that

$$H = H(x, p, t) \tag{12.45}$$

For this, let us first express the variations of H in its form (12.44)

$$\delta H = \sum_{k}^{n} (\dot{x}_k \, \delta p_k + p_k \, \delta \dot{x}_k - \frac{\partial L}{\partial x_k} \delta x_k - \frac{\partial L}{\partial \dot{x}_k} \delta \dot{x}_k) \qquad (12.46)$$

The definition (12.42) allows one to simplify the previous expression

$$\delta H = \sum_{k}^{n} (\dot{x}_k \, \delta p_k - \frac{\partial L}{\partial x_k} \delta x_k) \qquad (12.47)$$

When derived from equation (12.45), this same variation δH takes the form

$$\delta H = \sum_{k}^{n} (\frac{\partial H}{\partial x_k} \delta x_k + \frac{\partial H}{\partial p_k} \delta p_k) \qquad (12.48)$$

By comparing the two expressions above, one can write

$$\dot{x}_k = \frac{\partial H}{\partial p_k} \qquad k = 1, 2, \ldots, n \qquad (12.49)$$

$$-\frac{\partial L}{\partial x_k} = \frac{\partial H}{\partial x_k} \qquad k = 1, 2, \ldots, n \qquad (12.50)$$

Equations (12.49) and (12.50) have the transformation of Legendre as their only source. By using the equations for the dynamics (12.41), one obtains

$$\dot{p}_k = \frac{d}{dt} (\frac{\partial L}{\partial \dot{x}_k}) = \frac{\partial L}{\partial x_k} \qquad k = 1, 2, \ldots, n \qquad (12.51)$$

Equations (12.50) then become

$$\dot{p}_k = -\frac{\partial H}{\partial x_k} \qquad k = 1, 2, \ldots, n \qquad (12.52)$$

The two sets of equations (12.49) and (12.52), explicitly

$$\begin{cases} \dot{x}_k = \dfrac{\partial H}{\partial p_k} \\ \dot{p}_k = - \dfrac{\partial H}{\partial x_k} \end{cases} \qquad k = 1, 2, \ldots, n \tag{12.53}$$

constitute a system of $2n$ first order differential equations, called the **canonical equations of Hamilton.**

The function of Hamilton, or Hamiltonian, gives a complete description of the motion since all the differential equations of this motion can be deduced from it.

The advantage of the equations of Hamilton over those of Lagrange resides in the fact that the time derivatives only appear in the left-hand sides.

Let us recall that the $n$ first equations (12.53) result from the transformation of Legendre and from the definition of the Hamiltonian, whereas the $n$ following are the transcription of the laws of dynamics which govern motion.

In the presence of non-conservative forces, by differentiating the dissipation function $W$ of Lord Rayleigh, Hamilton's equations become

$$\begin{cases} x_k = \dfrac{\partial H}{\partial p_k} \\ \dot{p}_k = - \dfrac{\partial H}{\partial x_k} - \dfrac{\partial W}{\partial \dot{x}_k} \end{cases} \qquad k = 1, 2, \ldots, n \tag{12.54}$$

In the simplest and most common case where the kinetic energy is reduced to the positive definite quadratic form of the generalized velocities $\dot{x}_k$, the Hamiltonian has the value

$$H = T + V \tag{12.55}$$

When the kinetic and potential energies have the values (10.2) and (10.3) of chapter 10 respectively, the Lagrangian becomes

$$L = T - V = \frac{1}{2} \dot{\vec{x}}^T [M] \dot{\vec{x}} - \frac{1}{2} \vec{x}^T [K] \vec{x} \tag{12.56}$$

On the other hand the generalized momenta (12.42) are given by the expression

$$\vec{p} = [M] \dot{\vec{x}} \tag{12.57}$$

These relations allow one to calculate the Hamiltonian function (12.44)

$$H = \vec{p} \dot{\vec{x}} - L = \vec{p} \dot{\vec{x}} - \frac{1}{2} \dot{\vec{x}}^T [M] \dot{\vec{x}} + \frac{1}{2} \vec{x}^T [K] \vec{x}$$

then, by eliminating $\dot{\vec{x}}$ using (12.57)

$$H = \frac{1}{2} \vec{p}^T [M]^{-1} \vec{p} + \frac{1}{2} \vec{x}^T [K] \vec{x} \tag{12.58}$$

Taking account the value (10.4) of the dissipation function **W** , Hamilton's equations (12.54) are finally

$$\begin{cases} \dot{\vec{x}} = [M]^{-1} \vec{p} \\ \dot{\vec{p}} = - [K] \vec{x} - [C] \dot{\vec{x}} \end{cases} \tag{12.59}$$

In order to express the equations (12.59) in matrix form, it is preferable to avoid inversion of the mass matrix. One writes thus

$$\begin{cases} [M] \dot{\vec{x}} = \vec{p} \\ \dot{\vec{p}} + [C] \dot{\vec{x}} = - [K] \vec{x} \end{cases} \tag{12.60}$$

Using $\vec{p}$ and $\vec{x}$ on the one hand, and $\dot{\vec{p}}$ and $\dot{\vec{x}}$ on the other, one can then make some new vectors having 2n components, so as to put the preceding equations in the form

$$\begin{bmatrix} [0] & [M] \\ \lceil I \rfloor & [C] \end{bmatrix} \begin{Bmatrix} \dot{\vec{p}} \\ \dot{\vec{x}} \end{Bmatrix} = \begin{bmatrix} \lceil I \rfloor & [0] \\ [0] & -[K] \end{bmatrix} \begin{Bmatrix} \vec{p} \\ \vec{x} \end{Bmatrix} \tag{12.61}$$

In modal analysis, one generally prefers to deal with equations involving only the initial generalized coordinates $x_i$ and their derivatives. It suffices to transform the above system of 2n differential equations by the simple relations

$$\begin{Bmatrix} \vec{p} \\ \vec{x} \end{Bmatrix} = \begin{bmatrix} [M] & [0] \\ [0] & \lceil I \rfloor \end{bmatrix} \begin{Bmatrix} \dot{\vec{x}} \\ \vec{x} \end{Bmatrix} \tag{12.62}$$

then by differentiation

$$\begin{Bmatrix} \dot{\vec{p}} \\ \dot{\vec{x}} \end{Bmatrix} = \begin{bmatrix} [M] & [0] \\ [0] & \lceil I \rfloor \end{bmatrix} \begin{Bmatrix} \ddot{\vec{x}} \\ \dot{\vec{x}} \end{Bmatrix} \tag{12.63}$$

The system (12.61) takes the form

$$\begin{bmatrix} [0] & [M] \\ [M] & [C] \end{bmatrix} \begin{Bmatrix} \ddot{\vec{x}} \\ \dot{\vec{x}} \end{Bmatrix} = \begin{bmatrix} [M] & [0] \\ [0] & -[K] \end{bmatrix} \begin{Bmatrix} \dot{\vec{x}} \\ \vec{x} \end{Bmatrix} \tag{12.64}$$

The matrices of order 2n are symmetrical and are made up from elements which are themselves square and symmetric matrices.

One can see that by expanding the system (12.64), one obtains the two systems of n equations

$$\begin{cases} [M]\,\dot{\vec{x}} = [M]\,\dot{\vec{x}} \\ [M]\,\ddot{\vec{x}} + [C]\,\dot{\vec{x}} = -\,[K]\,\vec{x} \end{cases} \tag{12.65}$$

The first system of order n is trivial, while the second represents the equations for the dynamics of the physical system.

The transformation of a system of n second order differential equations into a system of 2n first order equations by the addition of a trivial system of order n has been developed by Frazer, Duncan and Collar [7], without reference to a Hamiltonian formulation of the problem. It is for this that one often encounters in the literature the term **Duncan transformation** for the transition from the system (12.1) to the system (12.64).

One sees immediately, by examining (12.64) or (12.65), that if the system is subject to some external forces $f_i(t)$ the force-vector of order 2n will be

$$\vec{p}^T = \{\vec{0}^T \ \vec{f}^T\} \tag{12.66}$$

In this expression, $\vec{0}^T$ and $\vec{f}^T$ are respectively the transpose of the null vector of order n and of the generalized force vector of equation (10.1).

$$\vec{f}^T = \{f_1 \ f_2 \ \dots \ f_n\} \tag{12.67}$$

## 12.6 Solution of the differential system

### 12.6.1 Change of coordinates - Phase space

Let us again take the system of differential equations (12.64) and let us introduce the following notation in order to simplify the writing :

$$[D] = \begin{bmatrix} [0] & [M] \\ [M] & [C] \end{bmatrix} \tag{12.68}$$

$$[G] = \begin{bmatrix} -[M] & [0] \\ [0] & [K] \end{bmatrix} \tag{12.69}$$

$$\begin{cases} \vec{y}^T = \{\dot{\vec{x}}^T \ \vec{x}^T\} \\ \dot{\vec{y}}^T = \{\ddot{\vec{x}}^T \ \dot{\vec{x}}^T\} \end{cases} \tag{12.70}$$

The differential equation of the system in free state becomes simply

$$[D]\ \dot{\vec{y}} + [G]\ y = 0 \tag{12.71}$$

Like the matrices of order n ([M] , [C] and [K]) , the matrices of order 2n ([D] and [G]) are square, real and symmetric.

Moreover, the mass matrix [M] being always positive definite, the matrix [D] is always invertible, its determinant being equal, apart from the sign, to the square of that of [M]

$$|D| = (-1)^n\ |M|^2 \tag{12.72}$$

By means of the obvious relation $[D]\ [D]^{-1} = [I] = [D]^{-1}[D]$ it is easy to calculate analytically this inverse matrix. One obtains

$$[D]^{-1} = \begin{bmatrix} -[M]^{-1}[C]\ [M]^{-1} & [M]^{-1} \\ [M]^{-1} & [0] \end{bmatrix} \tag{12.73}$$

The differential system (12.71), of the order and of dimension 2 n , can be solved in a way analogous to that used in section 11.3 for the system (11.1). The nature of the matrices being the same in the two cases, it is possible to decouple the system (12.71) by a change of variables

$$\vec{y} = [B]\ \vec{q} \tag{12.74}$$

The change of basis matrix [B] is here square, with constant but complex coefficients, and of dimension 2n. There are also 2n decoupled coordinates $q_i$. By differentiation one obtains

$$\dot{\vec{y}} = [B]\ \dot{\vec{q}} \tag{12.75}$$

By making this change of variables in the initial system and by pre-multiplying by $[B]^T$ in order to conserve the symmetry of the matrix products we get

$$[B]^T[D]\ [B]\ \dot{\vec{q}} + [B]^T[G]\ [B]\ \vec{q} = \vec{0} \tag{12.76}$$

The procedure of section 11.3 allows one to assert that there exists a matrix $[B]$ such that the products

$$[B]^T[D]\ [B] = \lceil D^\circ \rfloor \tag{12.77}$$

and

$$[B]^T[G]\ [B] = \lceil G^\circ \rfloor \tag{12.78}$$

are diagonal matrices. This matrix $[B]$ satisfies the condition

$$[B]^{-1}[D]^{-1}[G]\ [B] = \lceil \Delta \rfloor \tag{12.79}$$

One is thus brought back to an eigenvalue problem analogous to the one dealt with before.

### 12.6.2 Eigenvalue problem

The core of the system, of order 2n, is here

$$[F] = [D]^{-1}[G] \tag{12.80}$$

With this notation, the condition for diagonalization (12.79) simplifies

$$[B]^{-1}[F]\ [B] = \lceil \Delta \rfloor \tag{12.81}$$

By making the matrix product of (12.73) with (12.69), one finds, ($[I]$ being the unit matrix of order $n$) :

$$[F] = \begin{bmatrix} [M]^{-1}[C] & [M]^{-1}[K] \\ -\lceil I \rfloor & [0] \end{bmatrix} \tag{12.82}$$

The search for the eigenvalues $\delta_p$ which form the diagonal matrix $[\Delta]$ is made in the usual way by solution of the characteristic equation

$$| [F] - \delta_p \lceil I \rfloor = 0 | \tag{12.83}$$

This equation, in which $\lceil I \rfloor$ is the unit matrix of order $2n$, is identical to (12.35). By replacing $[F]$ by its value, it is written

$$\left| \begin{bmatrix} [M]^{-1}[C] & [M]^{-1}[K] \\ -\lceil I \rfloor & [0] \end{bmatrix} - \delta_p \lceil I \rfloor \right| = 0 \tag{12.84}$$

We have assumed (page 241) that all the roots are complex conjugates

$$\begin{cases} \delta_p = \lambda_p + j\,\omega_p \\ \delta_p^* = \lambda_p - j\,\omega_p \end{cases} \tag{12.85}$$

The $2n$ eigenvectors $\vec{B}_p$, which are linked to the $2n$ eigenvalues, are defined apart from a factor, and they constitute the change of basis matrix $[B]$. They can be obtained, either as solutions of the systems

$$[[F] - \delta_p \lceil I \rfloor]\, \vec{B}_p = \vec{0} \tag{12.86}$$

or as column vectors of the adjoint matrices of $[[F] - \delta \lceil I \rfloor]_p$. These eigenvectors are no longer real, as was the case before, but complex. It is easy to show that two conjugated eigenvectors correspond to two conjugated eigenvalues, since the terms of $[F]$ are real.

The system (12.76) being decoupled, we can write, taking account of the relations (12.77) and (12.78)

$$\lceil D^\circ \rfloor\, \dot{\vec{q}} + \lceil G^\circ \rfloor\, \vec{q} = \vec{0} \tag{12.87}$$

then, by pre-multiplying by $[D°]^{-1}$

$$\dot{\vec{q}} + [D°]^{-1}[G°]\,\vec{q} = \vec{0} \qquad (12.88)$$

Since

$$[D°]^{-1}[G°] = [\Delta] \qquad (12.89)$$

it finally gives

$$\dot{\vec{q}} + [\Delta]\,\vec{q} = \vec{0} \qquad (12.90)$$

12.6.3 General solution

The decoupled system (12.90), consists of $2n$ independent equations

$$\dot{q}_p + \delta_p\, q_p = 0 \qquad p = 1, 2, \ldots, 2n \qquad (12.91)$$

The integration of the equations is then immediate and gives the solutions

$$q_p = Q_p\, e^{-\delta_p t} \qquad p = 1, 2, \ldots, 2n \qquad (12.92)$$

The constants of integration $Q_p$ are determined by the initial conditions imposed on the system.

By coming back to the initial coordinates by means of relation (12.74) for the change of basis

$$\vec{y} = [B]\,\vec{q}$$

one can write

$$\vec{y} = \sum_{p}^{2n} \vec{B}_p\, q_p \qquad (12.93)$$

then, by replacing the $q_p$ by their values

$$\vec{y} = \sum_{p}^{2n} \vec{B}_p Q_p e^{-\delta_p t} \tag{12.94}$$

By construction of the system of order 2n, the components of the vector $\vec{y}$ satisfy the condition

$$y_i = \frac{dy_{i+n}}{dt} \qquad i = 1, 2, \ldots, n \tag{12.95}$$

This results in

$$B_{ip} = - \delta_p B_{i+n,p} \tag{12.96}$$

One can arrange the 2n solutions so that the conjugate of $\delta_p$ is placed in row $p + n$

$$\delta_p^* = \delta_{p+n} \tag{12.97}$$

The relation (12.94) becomes in this way

$$\vec{y} = \sum_{p}^{n} [\vec{B}_p Q_p e^{-\delta_p t} + \vec{B}_p^* Q_{p+n} e^{-\delta_p^* t}] \tag{12.98}$$

Since the vector $\vec{y}$ represents the displacements and velocities of the system, its elements are necessarily real quantities, which implies

$$(\vec{B}_p^* Q_{p+n}) = (\vec{B}_p Q_p)^*$$

and consequently

$$Q_{p+n} = Q_p^* \tag{12.99}$$

then, by introducing this result into (12.98)

$$\vec{y} = \sum_{p}^{n} [\vec{B}_p Q_p e^{-\delta_p t} + \vec{B}_p^* Q_p^* e^{-\delta_p^* t}] \tag{12.100}$$

Let us write the complex quantities in exponential form

$$\begin{cases} \vec{B}_p = \{\beta_{ep}\ e^{j\psi_{\ell p}}\} & \ell = 1,\ 2,\ \dots,\ 2\ n \\ \vec{B}^*_p = \{\beta_{ep}\ e^{-j\psi_{\ell p}}\} & \ell = 1,\ 2,\ \dots,\ 2\ n \end{cases} \tag{12.101}$$

$$\begin{cases} Q_p = \frac{1}{2}\ Y_p\ e^{j\varphi_p} \\ Q^*_p = \frac{1}{2}\ Y_p\ e^{-j\gamma_p} \end{cases} \tag{12.102}$$

Because of the specific form of the eigenvalues (12.85), this gives

$$\vec{y} = \sum_p^n \frac{1}{2}\ Y_p\ e^{-\lambda_p}[\{\beta_{ep}\ e^{-j(\omega_p t-\psi_{\ell p}-\varphi_p)}\} + \{\beta_{\ell p}\ e^{j(\omega_p t-\psi_{\ell p}-\varphi_p)}\}] \tag{12.103}$$

and, by introducing the harmonic solutions

$$\vec{y} = \sum_p^n e^{-\lambda_p t}\{Y_p\ \beta_{\ell p}\ \cos(\omega_p t-\psi_{\ell p}-\varphi_p)\} \qquad \ell = 1,\ 2,\ \dots,\ 2\ n \tag{12.104}$$

or

$$\vec{y} = \sum_p^n \{\beta_{\ell p}\ Y_p\ e^{-\lambda_p t}\ \cos(\omega_p t-\psi_{\ell p}-\varphi_p)\}$$

The vector $\vec{y}$ being made up of the velocities and displacements, the $n$ last components of (12.104) suffice to describe the motion of the system. Consequently, by modifying the notation as follows

$$i = \ell - n \qquad \text{with } i = 1,\ 2,\ \dots,\ n$$

$$X_p = Y_p$$

the equation of motion for the **dissipative free state with complex modes** becomes finally

$$\vec{x} = \sum_p^n \{\beta_{ip}\ X_p\ e^{-\lambda_p t}\ \cos(\omega_p t-\psi_{ip}-\varphi_p)\} \tag{12.105}$$

Comparison of (12.105) and (12.24) underscores the fact that when the damping of the system does not satisfy the Caughey condition there appear, for a given natural mode, not only different amplitudes $\beta_{ip}$ for each coordinate $x_i$ of the system, but also different phase shifts $\psi_{ip}$ .

A natural mode shape is then no longer, as in a conservative system, a static configuration of the system linked to an eigenvalue. It must be defined as a configuration **in phase space, linked to an eigenvalue.** This space is made up from the generalized displacements and the generalized velocities.

On the other hand, one calls the pure number defined as follows, the **modal damping factor** or **relative modal damping**

$$\eta_p = \frac{\mathrm{Re}(\delta_p)}{|\delta_p|} = \frac{\lambda_p}{\sqrt{\lambda_p^2 + \omega_p^2}} \tag{12.106}$$

It is equivalent to that of the elementary oscillator, which would have the value

$$\eta = \frac{\lambda}{\omega_0} = \frac{\lambda}{\sqrt{\lambda^2 + \omega_1^2}}$$

12.6.4 Orthogonality of the modal vectors · Normalization

The directions of the coordinates $y_i$ , that is to say those of the coordinates $\dot{x}_i$ and $x_i$, are linearly independant. They make up a basis for the phase space, of dimension $2n$ . The complex modal vectors $\vec{B}_p$ constitute another basis for this space.

As previously, the linear independance of the modal vectors is expressed by a relation of the form

$$\sum_p^{2n} \gamma_p \vec{B}_p \neq \vec{0} \tag{12.107}$$

where the $\gamma_p$ are some arbitrary complex constants. The orthogonality property of the modal vectors, necessary for the demonstration of (12.107), is also fulfilled in the case of a dissipative system. In effect, the diagonalization of [D] and [G] has led to the relations (12.77) and (12.78)

$$[B]^T[D]\,[B] = \lceil D^\circ \rfloor$$

$$[B]^T[G]\,[B] = \lceil G^\circ \rfloor$$

By introducing the column vectors $\vec{B}_p$ of [B], one can write, for the first of these relations :

$$\begin{bmatrix} \vec{B}_1^T \\ \vdots \\ \vec{B}_{2n}^T \end{bmatrix} [D] \; \{ \vec{B}_1 \; \dots \; \vec{B}_{2n} \} = \begin{bmatrix} d_1^o & & 0 \\ & \ddots & \\ 0 & & d_{2n}^o \end{bmatrix} \qquad (12.108)$$

By considering this term by term, one deduces from it the equalities ($\delta_{rs}$ being Kronecker's delta) :

$$\vec{B}_r^T[D]\,\vec{B}_s = \delta_{rs}\, d_r^\circ \qquad (12.109)$$

then, by doing likewise with the matrix [G] (12.110)

$$\vec{B}_r^T[G]\,\vec{B}_s = \delta_{rs}\, g_r^\circ$$

When $r \neq s$, the preceding relations are equal to zero and express the property, already encountered previously, called the **orthogonality of the modal vectors** or **orthogonality of the natural modes**. The essential difference from the relations (11.60) and (11.61), which are valid for a conservative system or a dissipative system satisfying the Caughey condition, is that the modal vectors above are complex and of dimension 2 n.

When $r = s$, one finds again the modal quantities, complex and conjugated in pairs, which make up the matrices $\lceil D^\circ \rfloor$ and $\lceil G^\circ \rfloor$, that is to say the terms $d_r^\circ$ and $g_r^\circ$. These are the counterparts of the modal masses and modal stiffnesses introduced in chapter 11. However, the nature of the matrices [D] and [G] no longer allows one to give them a simple physical meaning.

Since the eigenvectors are complex, the normalization giving a unit length to these vectors has hardly any meaning. In practice, one generally uses the normalization **which sets the terms of the matrix $[D^\circ]$ to unity**. It corresponds to the condition

$$[B]^T[D]\,[B] = \lceil I \rfloor \tag{12.111}$$

## 12.7 Response to an initial excitation in the general case

Let us consider a release with any initial conditions

$$\vec{x}(0) = \vec{X}_0$$

$$\dot{\vec{x}}(0) = \vec{V}_0$$

From the definition of the vector $\vec{y}$, of order 2n, the vector $\vec{Y}_0$ for the initial conditions is of the form

$$\vec{Y}_0 = \begin{Bmatrix} \vec{V}_0 \\ \vec{X}_0 \end{Bmatrix} \tag{12.112}$$

The general solution (12.94) allows one to write

$$\vec{Y}_0 = \sum_p^{2n} \vec{B}_p\, Q_p \tag{12.113}$$

The $Q_p$ can be determined by using the orthogonality relations (12.109). Let us pre-multiply the preceding relation by $\vec{B}_r^T[D]$

$$\vec{B}_r^T[D]\ \vec{Y}_0 = \sum_p^{2n} \vec{B}_r^T[D]\ \vec{B}_p\ Q_p$$

The terms of the sum are all zero, with the exception of those of index $r$ , that is to say

$$\vec{B}_r^T[D]\ \vec{Y}_0 = d_r^\circ\ Q_r$$

and consequently

$$Q_r = \frac{1}{d_r^\circ}\ \vec{B}_r^T[D]\ \vec{Y}_0 \qquad (12.114)$$

By transferring this expression into the general solution it gives

$$\vec{y} = \sum_p^{2n} \frac{1}{d_p^\circ}\ (\vec{B}_p^T[D]\ \vec{Y}_0)\ \vec{B}_p\ e^{-(\lambda_p + j\omega_p)t}$$

We had arranged the solutions in such a way that

$$\delta_{p+n} = \delta_p^* \qquad p = 1,\ 2,\ \ldots,\ n \qquad (12.115)$$

We shall have in the same way (12.116)

$$\vec{B}_{p+n} = \vec{B}_p^*$$

and

$$d_{p+n}^\circ = d_p^{\circ *}$$

By setting moreover

$$\vec{B}_p = \{\beta_{\ell p}\ e^{j\psi_{\ell p}}\}$$

the solution can be put in the form of a sum of n terms

$$\vec{y} = \sum_{p}^{n} [\frac{1}{d_p^\circ} (\{\beta_{\ell p}\ e^{j\psi_{\ell p}}\}^T [D]\ \vec{Y}_0)\ \{\beta_{\ell p}\ e^{j\psi_{\ell p}}\}\ e^{-(\lambda_p+j\omega_p)t} + \\ + \frac{1}{d_p^{\circ *}} (\{\beta_{\ell p}\ e^{-j\psi_{\ell p}}\}^T [D]\ \vec{Y}_0)\ \{\beta_{\ell p}\ e^{-j\psi_{\ell p}}\}\ e^{-(\lambda_p-j\omega_p)t}] \tag{12.117}$$

Let us simplify this result by modifying the notation for the constants

$$\begin{cases} \frac{1}{2} Y'_{0p} e^{j\varphi_{0p}} = \frac{1}{d_p^\circ} \{\beta_{\ell p}\ e^{j\psi_{\ell p}}\}^T [D]\ \vec{Y}_0 \\ \frac{1}{2} Y'_{0p} e^{-j\varphi_{0p}} = \frac{1}{d_p^{\circ *}} \{\beta_{\ell p}\ e^{-j\psi_{\ell p}}\}^T [D]\ \vec{Y}_0 \end{cases} \tag{12.118}$$

It gives in this way

$$\vec{y} = \sum_{p}^{n} \{\beta_{\ell p}\ Y'_{0p} e^{-\lambda_p t} \cos(\omega_p t - \psi_{\ell p} - \varphi_{0p})\} \tag{12.119}$$

One could have established this result more quickly using (12.114), once the coefficients $Q_p$ were determined. In effect, it was then possible to adopt directly the notation of equation (12.102).

By making the same change of index $i = \ell - n$ as before and by replacing $Y'_{0p}$ by $X'_{0p}$, the motion of the system can be described by the vector $\vec{x}$ consisting of only n components

$$\vec{x} = \sum_{p}^{n} \{\beta_{ip}\ X'_{0p} e^{-\lambda_p t} \cos(\omega_p t - \psi_{ip} - \varphi_{0p})\} \tag{12.120}$$

The vector $\vec{Y}_0$ of the initial conditions being always real, the general conditions which allow one to isolate the mode of the row r is written

$$\vec{Y}_0 = \gamma_r\ \vec{B}_r + \gamma_r^*\ \vec{B}_r^* \tag{12.121}$$

In this expression, $\gamma_r$ and $\gamma_r^*$ are two complex conjugate constants which one can put in the form

$$\left\{\begin{aligned} \gamma_r &= \frac{1}{2}\,Y'_{0r}\;e^{j\varphi_{0r}} \\ \gamma_r^* &= \frac{1}{2}\,Y'_{0r}\;e^{-j\varphi_{0r}} \end{aligned}\right. \tag{12.122}$$

By means of the exponential expressions (12.101) of the eigenvectors, the relation (12.121) becomes

$$\vec{Y}_0 = \{\beta_{\ell r}\;e^{j\psi_{\ell r}}\}\;Y'_{0r}\;e^{j\varphi_{0r}} + \{\beta_{\ell r}\;e^{-j\psi_{\ell r}}\}\;Y'_{0r}\;e^{-j\varphi_{0r}} \tag{12.123}$$

By introducing these initial conditions into the response of the system (12.117) and by using the orthogonality relations (12.109) one obtains

$$\begin{aligned} \vec{y} = \frac{1}{2}\,Y'_{0r}\;e^{j\varphi_{0r}}\{\beta_{\ell r}\;e^{j\psi_{\ell r}}\}\;e^{-(\lambda_r+j\omega_r)t} \\ + \frac{1}{2}\,Y'_{0r}\;e^{-j\varphi_{0r}}\{\beta_{\ell r}\;e^{-j\psi_{\ell r}}\}\;e^{-(\lambda_r-j\omega_r)t} \end{aligned} \tag{12.124}$$

This gives finally, by grouping the terms and by coming back to the harmonic functions

$$\vec{y} = Y'_{0r}\{\beta_{\ell r}\;e^{-\lambda_r t}\;\cos(\omega_r t-\psi_{\ell r}-\varphi_{0r})\} \tag{12.125}$$

As before, the vector $\vec{x}$ of the displacements is sufficient to describe the motion of the system

$$\vec{x} = X'_{0r}\{\beta_{ir}\;e^{-\lambda_r t}\;\cos(\omega_r t-\psi_{ir}-\varphi_{0r})\} \tag{12.126}$$

This result shows that in order to isolate a mode in a damped system, it is no longer sufficient to choose an initial static configuration. It is also necessary, because of the phase shifts $\psi_{ir}$, to impose a value determined by (n-1) of the velocities $\dot{x}_i$, only one of them - for example the first

$\dot{x}_1$ - can be zero. This condition makes it very difficult to isolate a mode in a real system.

In chapter 13, we are going to tackle a simple example (two degrees of freedom), in order to illustrate the concept of complex modes, so that the reader can clearly grasp the difference between real and complex modes.

## 12.8 Direct search for specific solutions

When the system of differential equations (12.71) is established, either by differentiating Hamilton's canonical equations, or by means of Duncan's transformation, their solution can also be tackled by means of a search for specific solutions for the vector $\vec{y}$. In effect, we had

$$[D]\,\dot{\vec{y}} + [G]\,\vec{y} = \vec{0}$$

$$\text{with} \quad \vec{y} = \begin{Bmatrix} \dot{\vec{x}} \\ \vec{x} \end{Bmatrix} \quad \text{and} \quad \dot{\vec{y}} = \begin{Bmatrix} \ddot{\vec{x}} \\ \dot{\vec{x}} \end{Bmatrix} \tag{12.127}$$

Let us therefore search for solutions of $\vec{y}$ of the form

$$\vec{y} - \vec{B}_p\, e^{-\delta_p t} \tag{12.128}$$

By introducing these into the equation of the system, it gives

$$[-\,\delta_p[D] + [G]]\,\vec{B}_p\, e^{-\delta_p t} = \vec{0}$$

By simplifying this equation using the non-zero quantity $e^{-\delta_p t}$, and by adopting the convention of (12.80)

$$[F] = [D]^{-1}[G]$$

then by seeking the existence condition for solutions which are not all zero, one again finds the characteristic equation (12.83)

$$\left| [F] - \delta_p [I] \right| = 0$$

On the other hand, the eigenvectors are obtained by solution of the 2 n homogeneous systems (12.86)

$$[[F] - \delta_p [I]] \vec{B}_p = \vec{0}$$

## 12.9 Another form of the characteristic equation

If one disposes of the flexibility matrix $[\alpha]$ , which is the inverse of the stiffness matrix $[K]$ , it is convenient to write the matrix differential equation (12.1) of the system

$$[\alpha]\ [M]\ \ddot{\vec{x}} + [\alpha]\ [C]\ \dot{\vec{x}} + \vec{x} = \vec{0} \tag{12.129}$$

In order to obtain a differential system of order $2n$ , it is necessary to add to this equation the following trivial equation, in accordance with Duncan's method

$$[\alpha]\ [M]\ \dot{\vec{x}} - [\alpha]\ [M]\ \dot{\vec{x}} = 0$$

The two preceding equations can be grouped as follows

$$\begin{bmatrix} [0] & [\alpha]\ [M] \\ [\alpha]\ [M] & [\alpha]\ [C] \end{bmatrix} \begin{Bmatrix} \ddot{\vec{x}} \\ \dot{\vec{x}} \end{Bmatrix} + \begin{bmatrix} -[\alpha]\ [M] & [0] \\ [0] & [I] \end{bmatrix} \begin{Bmatrix} \dot{\vec{x}} \\ \vec{x} \end{Bmatrix} = \begin{Bmatrix} \vec{0} \\ \vec{0} \end{Bmatrix} \tag{12.130}$$

Apart from certain exceptional circumstances, which we shall not consider here, the matrix $[\alpha]$ is positive definite, and therefore invertible

Consequently one can invert the above diagonal matrix

$$\begin{bmatrix} -[\alpha]\,[M] & [0] \\ [0] & [I] \end{bmatrix}^{-1} = \begin{bmatrix} -[M]^{-1}[\alpha]^{-1} & [0] \\ [0] & [I] \end{bmatrix} \tag{12.131}$$

By pre-multiplying the system (12.130) by the inverse matrix (12.131), then by adopting the notation

$$[U] = \begin{bmatrix} [0] & [I] \\ [\alpha]\,[M] & [\alpha]\,[C] \end{bmatrix} \tag{12.132}$$

one obtains, taking account of (12.127)

$$[U]\,\dot{\vec{y}} + \vec{y} = \vec{0} \tag{12.133}$$

As before, one seeks specific solutions

$$\vec{y} = \vec{B}\ e^{-\delta_p t} \tag{12.134}$$

With the notation

$$w_p = \frac{1}{\delta_p} \tag{12.135}$$

the procedure already described leads to the solution of the characteristic equation

$$|[U] - w_p[I]| = 0 \tag{12.136}$$

The eigenvectors are given by the 2n homogeneous relations

$$[[U] - w_p[I]]\ \mathbf{B}_p = \mathbf{0}$$

# CHAPTER 13 EXAMPLE OF VISUALIZATION OF COMPLEX NATURAL MODES

## 13.1 Description of the system

A natural mode remains a relatively abstract idea because of the nature and the dimension of the space considered. In effect, the $n$ initial generalized coordinates $x_i$ in general have different physical dimensions (lengths, angles, etc.), so it is illusory to hope to find a physical interpretation of the space of which they constitute a basis.

Nevertheless, when all the coordinates $x_i$ are lengths and their number is less than or equal to three, the space defined is easily interpretable; it is either a straight line, a plane or a euclidian space depending on whether $n$ has the value one, two or three respectively.

Given the objective of simplifying as much as possible the mathematical formulation and presentation of the modes, we have chosen a system consisting only of a point mass which moves in a plane and which is kept in its equilibrium position by $r$ springs of stiffnesses $k_j$ . Moreover, $t$ viscous resistances with constants $c_j$ act on the mass. Such a system has two degrees of freedom (figure 13.1).

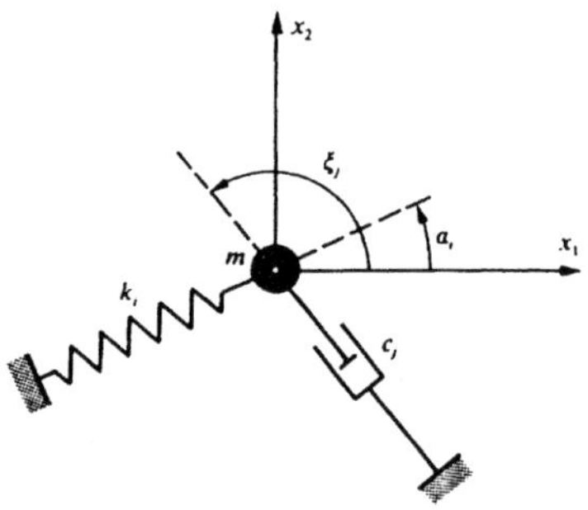


Fig. 13.1 Point mass attached to $r$ springs and $t$ viscous resistances

The lines of action of the springs $k_i$ are specified by the angles $\alpha_i$ , and those of the resistances by the angles $c_j$ . So as to retain the linearity of the problem, we assume that the elastic and dissipative elements are sufficiently long in order that the angles $\alpha_i$ and $\xi_j$ can be considered as constants at the time of the displacement of the mass.

## 13.2 Energetic form · Differential equation

It is convenient here to use the equations of Lagrange (10.11) in order to establish the matrix differential equation of motion of the mass

The **kinetic energy** has the elementary form

$$T = \frac{1}{2} m \, (\dot{x}_1^2 + \dot{x}_2^2) \tag{13.1}$$

The mass matrix which follows from it is then simply

$$[M] = \begin{bmatrix} m & 0 \\ 0 & m \end{bmatrix} = m \, [I] \tag{13.2}$$

In order to calculate the **potential energy,** let us consider any displacement $\vec{OO'}$ of the mass m (figure 13.2).

Fig. 13.2 Extension of the spring $k_i$ for the displacement $\vec{OO'}$ of the mass

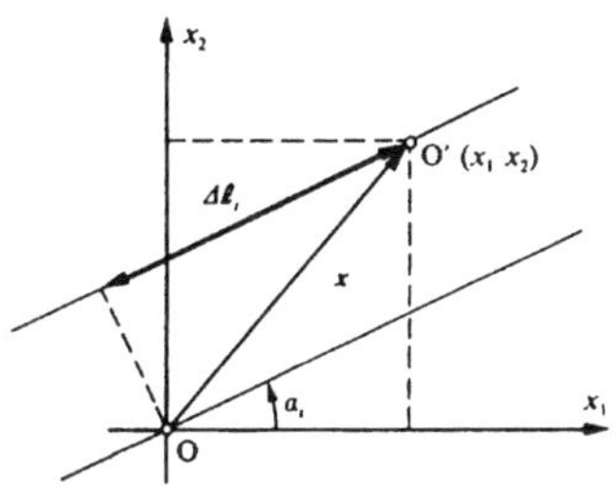

For this displacement, whose line of action is specified by the angle $k_i$ , which is assumed to be constant, the spring $\alpha_i$ is stretched by the amount $\Delta\ell_i$

$$\Delta\ell_i = x_1 \cos \alpha_i + x_2 \sin \alpha_i \tag{13.3}$$

The potential energy of the system is given by the sum

$$V = \frac{1}{2} \sum_i^r k_i \, \Delta\ell_i^2 \tag{13.4}$$

Its partial derivatives are given by the expression

$$\frac{\partial V}{\partial x_s} = \sum_i^r k_i \, \Delta\ell_i \, \frac{\partial \Delta\ell_i}{\partial x_s} \qquad s = 1,\ 2 \tag{13.5}$$

That being the case, one derives the stiffness matrix from it easily

$$[K] = \sum_i^r k_i \begin{bmatrix} \cos^2 \alpha_i & \sin \alpha_i \cos \alpha_i \\ \sin \alpha_i \cos \alpha_i & \sin^2 \alpha_i \end{bmatrix} \tag{13.6}$$

Although each matrix of index $i$ is singular, their sum is not, and the stiffness matrix is regular, with the exception of certain specific configurations which are not of interest in the present study.

By analogy with the result established for the elongation of the springs, the velocity in the direction $\xi_j$ of a viscous resistance is simply

$$\Delta\dot{\ell}_j = \dot{x}_1 \cos \xi_j + \dot{x}_2 \sin \xi_j \tag{13.7}$$

The **function of dissipation,** that is to say the **dissipated total half-power,** has the value

$$W = \frac{1}{2} \sum_{j}^{t} c_j \, \Delta\dot{\ell}_j^2 \tag{13.8}$$

The partial derivatives have the same form as before, that is to say

$$\frac{\partial W}{\partial \dot{x}_s} = \sum_{j}^{t} c_j \, \Delta\dot{\ell}_j \frac{\partial \Delta\dot{\ell}_j}{\partial \dot{x}_s} \qquad s = 1,\ 2 \tag{13.9}$$

The damping matrix therefore has the same form as that of the stiffness matrix

$$[C] = \sum_{j}^{t} c_j \begin{bmatrix} \cos^2 \xi_j & \sin \xi_j \cos \xi_j \\ \sin \xi_j \cos \xi_j & \sin^2 \xi_j \end{bmatrix} \tag{13.10}$$

By referring to the vector of displacements by $\vec{x}$ the **matrix differential equation** of the system is of the form (12.1)

$$[M]\,\ddot{\vec{x}} + [C]\,\dot{\vec{x}} + [K]\,\vec{x} = \vec{0}$$

In our specific case, since $[M] = m\,[I]$, the Caughey condition (13.11) expressed by (12.11) simplifies to

$$[C]\,[K] = [K]\,[C]$$

After having established that this condition is not satisfied, let us come back to the solution of the differential equation, given by (12.104), with $n = 2$ and $\ell = 1,\ 2,\ 3,\ 4$

$$\vec{y} = \{\beta_{\ell 1}\, Y_1\, e^{-\lambda_1 t} \cos(\omega_1 t - \psi_{\ell 1} - \varphi_1)\} + \{\beta_{\ell 2}\, Y_2\, e^{-\lambda_2 t} \cos(\omega_2 t - \psi_{\ell 2} - \varphi_2)\} \tag{13.12}$$

Let us recall that the vector $\vec{y}$ is defined by

$$\vec{y}^T = \{\dot{\vec{x}}^T \; \vec{x}^T\} = \{\dot{x}_1 \; \dot{x}_2 \; x_1 \; x_2\}$$

Finally, the motion of the mass is described by equation (12.105)

$$\begin{cases} x_1 = \beta_{11} \, X_1 \, e^{-\lambda_1 t} \cos(\omega_1 t - \psi_{11} - \varphi_1) + \beta_{12} \, X_2 \, e^{-\lambda_2 t} \cos(\omega_2 t - \psi_{12} - \varphi_2) \\ x_2 = \beta_{21} \, X_1 \, e^{-\lambda_1 t} \cos(\omega_1 t - \psi_{21} - \varphi_1) + \beta_{22} \, X_2 \, e^{-\lambda_2 t} \cos(\omega_2 t - \psi_{22} - \varphi_2) \end{cases} \quad (13.13)$$

## 13.3 Isolation of a mode

### 13.3.1 General case

The analytic condition for isolation of a mode is self-evident. It is sufficient, in the equation above, to set the amplitude $X_r$ to zero for the mode that one wants to eliminate. It then only remains to fix the two initial conditions of the remaining mode which becomes in this way

$$\begin{cases} x_1 = \beta_{1p} \, X_p \, e^{-\lambda_p t} \cos(\omega_p t - \psi_{1p} - \varphi_p) \\ x_2 = \beta_{2p} \, X_p \, e^{-\lambda_p t} \cos(\omega_p t - \psi_{2p} - \varphi_p) \end{cases} \qquad p = 1,2 \quad (13.14)$$

By analogy with the release of an elementary oscillator with given displacement and zero velocity, let us impose the following initial conditions.

$$x_1(0) = X_0 \qquad \text{et} \qquad \dot{x}_1(0) = 0 \quad (13.15)$$

In order to simplify the expression for $X_p$ and $\varphi_p$, we can choose the normalization of the eigenvectors as follows (13.16)

$$\beta_{1p} = 1$$

The quantities $\beta_{2p}$ and $\psi_{2p}$ can be replaced by $\beta_p$ and $\psi_p$, with (13.17)

$$\beta_p = \beta_{2p}/\beta_{1p} \qquad \text{and} \qquad \psi_p = \psi_{2p} - \psi_{1p}$$

The remaining mode is then described by the new relations

$$\begin{cases} x_1 = X_p\, e^{-\lambda_p t} \cos(\omega_p t - \varphi_p) \\ x_2 = \beta_p\, X_p\, e^{-\lambda_p t} \cos(\omega_p t - \psi_p - \varphi_p) \end{cases} \tag{13.18}$$

By introducing the initial conditions chosen in these equations one determines

$$X_p = \frac{X_0}{\cos \varphi_p} \quad \text{and} \quad \mathrm{tg}\, \varphi_p = \frac{\lambda_p}{\omega_p} \tag{13.19}$$

The initial conditions associated with the displacement $x_2$ can then be calculated

$$\begin{cases} x_2(0) = \beta_p\, X_0 (\cos \psi_p - \dfrac{\lambda_p}{\omega_p} \sin \psi_p) \\ \dot{x}_2(0) = \beta_p\, X_0 (- \lambda_p \dfrac{\cos(\varphi_p + \psi_p)}{\cos \varphi_p} + \omega_p \dfrac{\sin(\varphi_p + \psi_p)}{\cos \varphi_p}) \end{cases}$$

by expanding the trigonometric expressions of the second equation above, one obtains more simply

$$\begin{cases} x_2(0) = \beta_p\, X_0 (\cos \psi_p - \dfrac{\lambda_p}{\omega_p} \sin \psi_p) \\ \dot{x}_2(0) = \beta_p\, X_0 \dfrac{\lambda_p^2 + \omega_p^2}{\omega_p} \sin \psi_p \end{cases} \tag{13.20}$$

The initial conditions making possible the isolation of a mode are therefore fixed by the relations (13.15) and (13.20). One sees that it is not only necessary to bring the point mass to a given point in the plane $0\ x_1\ x_2$, but also to impose an initial velocity $\dot{x}_2(0)$ on it.

In practice that comes down to using a calibrated hammer blow to give the mass an impulse in the direction $x_2$, of magnitude

$$I = m\,\dot{x}_2(0) = m\,\beta_p\,X_0\,\frac{\lambda_p^2 + \omega_p^2}{\omega_p}\sin\psi_p \tag{13.21}$$

The natural mode retained is then governed by the equations

$$\left\{\begin{aligned} x_1 &= \frac{X_0}{\cos\varphi_p}\,e^{-\lambda_p t}\cos(\omega_p t-\varphi_p)\\ x_2 &= \beta_p\,\frac{X_0}{\cos\varphi_p}\,e^{-\lambda_p t}\cos(\omega_p t-\psi_p-\varphi_p)\end{aligned}\right. \tag{13.22}$$

These equations, which are of the parametric type, show that the trajectory of the mass corresponding to a natural mode is an **elliptical spiral** described in the plane $0\,x_1\,x_2$ of the system.

13.3.2 Principal axes of the trajectory

By definition, the principal axes cut the elliptical spiral orthogonally. That comes back to saying that the vector of displacements $\vec{x}$ and the vector of velocities $\dot{\vec{x}}$ have a zero scalar product at the intersection points between these axes and the trajectory

$$\vec{x}^T\,\dot{\vec{x}} = 0$$

In what follows, it is convenient to write this product in the form

$$1 + \frac{x_2\,\dot{x}_2}{x_1\,\dot{x}_1} = 0 \tag{13.23}$$

Differentiation of equations (13.22) gives

$$\left\{\begin{aligned} \dot{x}_1 &= -\frac{X_0}{\cos\varphi_p}\,e^{-\lambda_p t}\,(\lambda_p\cos(\omega_p t-\varphi_p) + \omega_p\sin(\omega_p t-\varphi_p))\\ \dot{x}_2 &= -\beta_p\,\frac{X_0}{\cos\varphi_p}\,e^{-\lambda_p t}\,(\lambda_p\cos(\omega_p t-\psi_p-\varphi_p) + \omega_p\sin(\omega_p t-\psi_p-\varphi_p))\end{aligned}\right. \tag{13.24}$$

By adopting the notation $u = \omega t - \varphi$, the condition (13.23) becomes

$$1 + \beta_p^2 \frac{\lambda_p \cos^2(u_p - \psi_p) + \omega_p \sin(u_p - \psi_p) \cos(u_p - \psi_p)}{\lambda_p \cos^2 u_p + \omega_p \sin u_p \cos u_p} = 0$$

then, after some trigonometric transformations,

$$1 + \beta_p^2 \frac{\operatorname{tg} \varphi_p[1 - \cos 2(\omega_p t - \psi_p)] + \sin 2(\omega_p t - \psi_p)}{\operatorname{tg} \varphi_p[1 - \cos 2\,\omega_p t] + \sin 2\,\omega_p t} = 0 \qquad (13.25)$$

This equation has two series of solutions

$$\begin{cases} t'_p = t'_{p0} + \gamma' \dfrac{\pi}{\omega_p} & \gamma' = 1, 2, \ldots \\ t''_p = t''_{p0} + \gamma'' \dfrac{\pi}{\omega_p} & \gamma'' = 1, 2, \ldots \end{cases} \qquad (13.26)$$

which correspond to the two angles $\theta'_p$ and $\theta''_p$ respectively, which define the directions of the principal axes of the trajectory. These angles, which we shall call the principal directions in what follows, are determined by their tangents

$$\begin{cases} \operatorname{tg} \theta'_p = \left.\dfrac{x_2}{x_1}\right|_{t = t'_p} \\ \operatorname{tg} \theta''_p = \left.\dfrac{x_2}{x_1}\right|_{t = t''_p} \end{cases} \qquad (13.27)$$

The principal directions of an actual natural mode are not themselves orthogonal, as we shall see in the numerical example dealt with below; nor are they orthogonal to the principal directions of the other natural mode either.

### 13.3.3 Conservative system

So as to allow the comparison of certain results, let us go back to the system of figure 13.1, but assume that it is without viscous resistances. The matrices [M] and [K] remain unchanged while the damping matrix is exactly zero. The eigenvalues are purely imaginary and the eigenvectors are real; consequently $\lambda_p = 0$ and $\psi_p = 0$ .
In order to isolate a mode, let us retain the initial conditions (13.15), that is to say

$$x_1(0) = X_0 \qquad \text{and} \qquad \dot{x}_1(0) = 0$$

The initial conditions (13.20) simplify to

$$x_2(0) = \beta_p X_0 \qquad \text{and} \qquad \dot{x}_2(0) = 0 \tag{13.28}$$

The two initial velocities being zero, it is a matter this time of a simple release of the mass, from a specific point in the plane $x_1\ x_2$ . The equations (13.22) of the conserved natural mode become, for the conservative system.

$$\begin{cases} x_1 = \dfrac{X_0}{\cos\varphi_p} \cos(\omega_p t - \varphi_p) \\ \\ x_2 = \beta_p \dfrac{X_0}{\cos\varphi_p} \cos(\omega_p t - \varphi_p) \end{cases} \tag{13.29}$$

This result shows that the trajectory of the mass is a segment of the straight line equation

$$x_2 = \beta_p\, x_1$$

In this specific case, the eigenvectors are (13.30)

$$\vec{\beta}_1 = \begin{Bmatrix} 1 \\ \beta_1 \end{Bmatrix} \qquad \text{and} \qquad \vec{\beta}_2 = \begin{Bmatrix} 1 \\ \beta_2 \end{Bmatrix}$$

Their orthogonality takes the form of a scalar product (11.60) since $[M] = m\,[I]$ . It gives therefore

$$\vec{\beta}_1^T \cdot \vec{\beta}_2 = 0 \quad => \quad 1 + \beta_1\,\beta_2 = 0$$

so that also

$$\beta_1 \cdot \beta_2 = -1 \tag{13.31}$$

Thus, the trajectories corresponding to the two natural modes are two segments of perpendicular straight lines.

## 13.4 Numerical examples

### 13.4.1 Equations of motion

So as to illustrate the concepts developed in this chapter, let us choose a numerical example consisting of three stiffnesses and three resistances

$m = 3$ kg

| | | |
|---|---|---|
| $k_1 = 700$ N/m | $k_2 = 1100$ N/m | $k_3 = 1300$ N/m |
| $\alpha_1 = 19$ ° | $\alpha_2 = 152$ ° | $\alpha_3 = 262$ ° |
| $c_1 = 15$ kg/s | $c_2 = 21$ kg/s | $c_3 = 13$ kg/s |
| $\xi_1 = 28$ ° | $\xi_2 = 160$ ° | $\xi_3 = 300$ ° |

With these values, the mass, stiffness, and damping matrices become respectively

$$[M] = \begin{bmatrix} 3 & 0 \\ 0 & 3 \end{bmatrix} \quad [K] = \begin{bmatrix} 1509 & -61.32 \\ -61.32 & 1591 \end{bmatrix} \quad [C] = \begin{bmatrix} 33.49 & -6.161 \\ -6.161 & 15.51 \end{bmatrix} \tag{13.32}$$

The core matrix $[F]$ , of order $2\,n$ is easily calculated (in order to simplify the writing, the first four significant figures only are shown below)

$$[F] = \begin{bmatrix} 11.16 & -2.054 & 502.8 & -20.44 \\ -2.054 & 5.171 & -20.44 & 530.5 \\ -1 & 0 & 0 & 0 \\ 0 & -1 & 0 & 0 \end{bmatrix} \tag{13.33}$$

The solutions of the characteristic equation (12.83) have the values

$$\begin{cases} \delta_1 = 2.350 + j\ 22.62 \\ \delta_2 = 5.817 + j\ 21.93 \\ \delta_3 = 2.350 - j\ 22.62 = \delta_1^* \\ \delta_4 = 5.817 - j\ 21.93 = \delta_2^* \end{cases} \tag{13.34}$$

and the matrix [B] of the modal vectors becomes :

$$[B] = \begin{bmatrix} 22.74\ e^{-j1.674} & 22.69\ e^{-j1.830} & 22.74\ e^{+j1.674} & 22.69\ e^{+j1.830} \\ 69.22\ e^{-j2.1990} & 6.870\ e^{+j1.857} & 69.22\ e^{+j2.199} & 6.870\ e^{-j1.857} \\ 1 & 1 & 1 & 1 \\ 3.044\ e^{-j0.5243} & 0.3028\ e^{-j2.596} & 3.044\ e^{j0.5243} & 0.3028\ e^{j2.596} \end{bmatrix} \tag{13.35}$$

The general motion of the mass is given by (13.13) :

$$\begin{cases} x_1 = X_1\ e^{-2.35t}\cos(22.62t - \varphi_1) + \\ \quad + X_2\ e^{-5.817t}\cos(21.93t - \varphi_2) \\ x_2 = 3.044\ X_1\ e^{-2.35t}\cos(22.62t + 0.5243 - \varphi_1) + \\ \quad + 0.3028\ X_2\ e^{-5.817t}\cos(21.93t + 2.596 - \varphi_2) \end{cases} \tag{13.36}$$

The modal damping factors are respectively using (12.106)

$$\eta_1 = 0.1033 \qquad \eta_2 = 0.2563$$

13.4.2 Isolation of the first mode

One chooses the initial conditions (13.15), that is to say $x_1(0) = X_0$ and $\dot{x}_1(0) = 0$ then one calculates, using (13.19)

$$\varphi_1 = \text{arctg}\ \frac{\lambda_1}{\omega_1} = \frac{2.350}{22.62} = 0.1035 = \pi/2 - 1.467$$

$$\frac{1}{\cos\varphi_1} = 1.0054$$

The relations (13.20) allow one next to determine

$$x_2(0) = 2.793 \qquad \dot{x}_2(0) = -34.84$$

The principal directions of the mode, obtained by solving equation (13.25), have the value

$$\theta'_1 = 74.66\,°$$
$$\theta''_1 = -17.19\,°$$

Finally, the first isolated mode is given by the equations

$$\begin{cases} x_1/X_0 = 1.005\, e^{-2.35t} \cos(22.62\, t - 0.1035) \\ x_2/X_0 = 3.060\, e^{-2,35t} \cos(22.62\, t + 0.4208) \end{cases} \qquad (13.37)$$

The trajectory described by the mass is an elliptical spiral shown in figure 13.3.

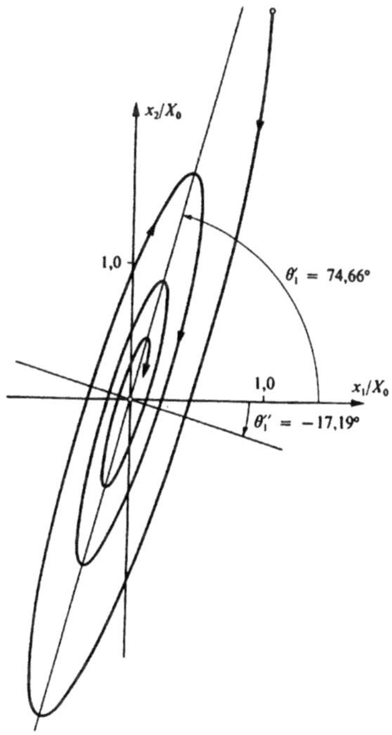


Fig. 13.3 Trajectory of the mass corresponding to the first complex mode of the system of figure 13.1

13.4.3 Isolation of the second mode

By proceeding in an analogous way, one obtains the following results for the second mode.

$$\varphi_2 = \text{Arc tg}\,\frac{\lambda_2}{\omega_2} = \frac{5.817}{21.93} = 0.2592 = \pi/2 - 1.3116$$

$$\frac{1}{\cos\varphi_2} = 1.0346$$

$$x_1(0) = X_0 \qquad \dot{x}_1(0) = 0$$

$$x_2(0) = 0.2171\ X_0 \qquad \dot{x}_2(0) = -\ 3.688\ X_0$$

The modal directions have the values

$$\theta_2' = -\ 12.55\ °$$
$$\theta_2'' = \ 72.88\ °$$

The equations governing the second isolated mode are as follows

$$\begin{cases} x_1/X_0 = 1.035\ e^{-5.817t}\cos(21.93\ t - 0.2592) \\ x_2/X_0 = 0.3132\ e^{-5.817t}\cos(21.93\ t + 2.337) \end{cases} \qquad (13.38)$$

Figure 13.4 shows the elliptical spiral corresponding to the second mode.

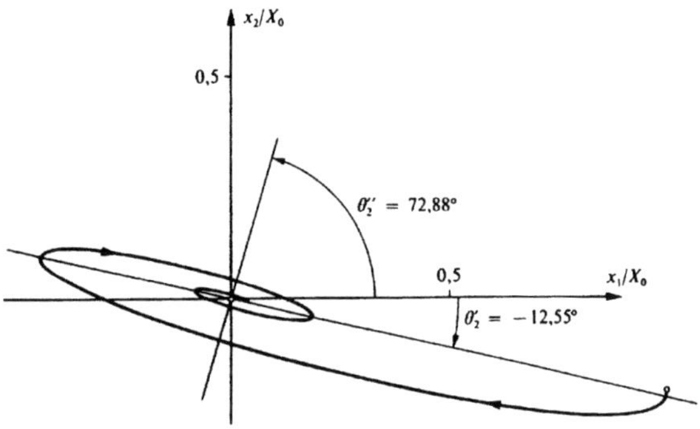


Fig. 13.4 Trajectory of the mass corresponding to the second complex mode of the system of figure 13.1

13.4.4 Conservative system

When the resistances $c_j$ are zero, the system is reduced to

$$[M]\,\ddot{\vec{x}} + [K]\,\vec{x} = \vec{0}$$

The mass and stiffness matrices remain defined by the relations (13.2) and (13.6) respectively. With the numerical values chosen, one obtains the following eigenvalues

$$\begin{cases} \delta_1 = \quad j\,\omega_1 = \quad j\ 22.18 \\ \delta_2 = \quad j\,\omega_2 = \quad j\ 23.27 \\ \delta_3 = -\,j\,\omega_1 = -\,j\ 22.18 \\ \delta_4 = -\,j\,\omega_2 = -\,j\ 23.27 \end{cases} \tag{13.39}$$

Since the modes are real, the changes of basis matrix $[B]$, of order $2\,n$, is written

$$[B] = \begin{bmatrix} 22.18\ e^{-j\pi/2} & 23.27\ e^{-j\pi/2} & 22.18\ e^{j\pi/2} & 23.27\ e^{j\pi/2} \\ 11.78\ e^{-j\pi/2} & 43.82\ e^{j\pi/2} & 11.78\ e^{j\pi/2} & 43.82\ e^{-j\pi/2} \\ 1 & 1 & 1 & 1 \\ 0.5310 & -\ 1.883 & 0.5310 & -\ 1.833 \end{bmatrix} \tag{13.40}$$

The motion is then described by the general equations

$$\begin{cases} x_1 = X_1 \cos(22.18\ t - \varphi_1) + X_2 \cos(23.27\ t - \varphi_2) \\ x_2 = 0.5310\ X_1 \cos(22.18\ t - \varphi_1) - 1.883\ X_2 \cos(23.27\ t - \varphi_2) \end{cases} \tag{13.41}$$

**Isolation of the first mode**

In this specific case, the initial conditions (13.15) lead to $\varphi_1 = 0$ and consequently

$$x_1(0) = X_0 \qquad \dot{x}_1(0) = 0$$

$$x_2(0) = 0.5310\ X_0 \qquad \dot{x}_2(0) = 0$$

The first mode is governed by the equations

$$\begin{cases} x_1/X_0 = \cos(22.18\ t) \\ x_2/X_0 = 0.5310 \cos(22.18\ t) \end{cases} \tag{13.42}$$

The corresponding trajectory is a segment of the straight line

$$x_2 = 0.5310\ x_1$$

which makes an angle with the axis $x_1$ which has the value

$$\theta_1 = \text{Arc tg}\ (0.5310) = 27.97\ °$$

**Isolation of the second mode**

As for the first mode, it gives $\varphi_2 = 0$ and the initial conditions are written

$$\begin{cases} x_1(0) = X_0 & \dot{x}_1(0) = 0 \\ x_2(0) = -\ 1.883\ X_0 & \dot{x}_2(0) = 0 \end{cases}$$

The equations of motion are then

$$\begin{aligned} x_1/X_0 &= \cos(23.27\ t) \\ x_2/X_0 &= -\ 1.883 \cos(23.27\ t) \end{aligned} \tag{13.43}$$

The trajectory is a segment of the straight line

$$x_2 = -\ 1.883\ x_1$$

making an angle $\theta_2$ with the axis $x_1$

$$\theta_2 = \text{Arc tg}(-1.883) = -\ 62.03\ °$$

The theoretical considerations of paragraph 13.3 have shown that the orthogonality of the modes implies, in this particular case, the perpendicularity of the modal straight lines. This is confirmed by the numerical results above, which are shown in figure 13.5.

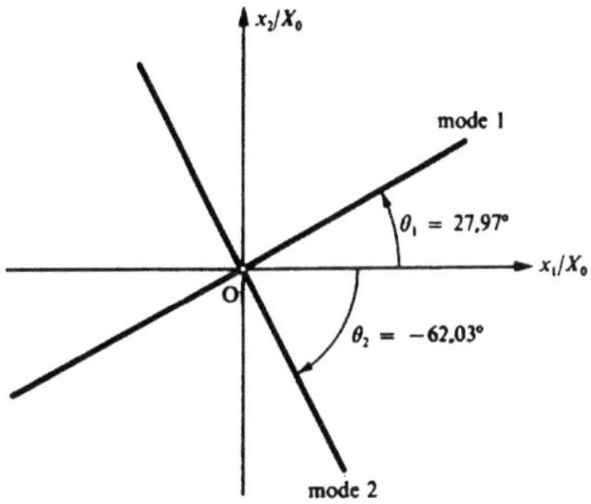


Fig. 13.5 Trajectories of the mass corresponding to the two real natural modes of the system of figure 13.1 without damping

## 13.5 Summary and comments

In this chapter, we have studied the behaviour of a coplanar system consisting of a point mass attached to any number of springs and viscous resistances (discrete linear elements). Such a system has two degrees of freedom, referred to by $x_1$ and $x_2$ . The results of this study can be summarized and commented upon as follows.

- When the system is dissipative, the two natural modes, known as complex modes, correspond to the trajectories of the mass which are two elliptical spirals. The axes of an actual spiral are not orthogonal, neither to themselves, nor to those of the other spiral. In order to isolate a vibratory mode, it is necessary to release the mass from a specific point in the plane $x_1 x_2$ , and also impose a certain initial velocity (for example according to $x_2$ as we have done it in this chapter).

- When the system is conservative (all the resistances are zero) the two modal trajectories are two segments of orthogonal straight lines. A mode can be isolated by releasing the mass, at zero initial velocity, from a specified point in the plane $x_1 x_2$

· The directions of the trajectories of the mass of a conservative system are the principle directions of the product $[M]^{-1}[K]$. As for the principle directions of the product $[M]^{-1}[C]$ they have the values $\theta_1 = 72{,}79°$, $\theta_2 = -17{,}21°$, in the numerical example chosen.

· The existence of a damping not respecting the Caughey condition leads to two consequences; on one hand the rectilinear trajectories are transformed into elliptical spirals, and on the other hand the principal directions of the complex modes are situated between the principal directions of $[M]^{-1}[K]$ and those of $[M]^{-1}[C]$. They become closer to the latter as the damping becomes more important (figure 13.6).

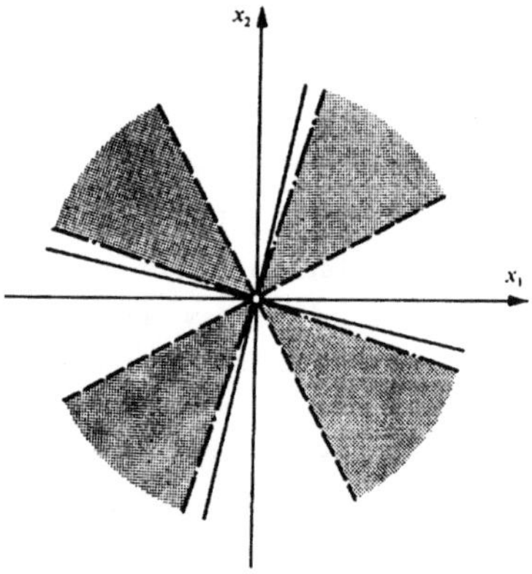


Fig. 13.6 Principal directions for the system of figure 13.1

: complex modes
: $[M]^{-1}[K]$
: $[M]^{-1}[C]$

· Therefore, the Caughey condition clearly expresses the congruence of the principal directions of $[M]^{-1}[K]$ with those of $[M]^{-1}[C]$.

# CHAPTER 14 FORCED STATE OF THE GENERALIZED OSCILLATOR

## 14.1 Introduction

For mechanical systems having many degrees of freedom, it is illusory to expect to undertake a systematic analytical study of the responses to the different types of excitation - that is to say, of the forced state - as we have done for the elementary oscillator, because of the very great diversity of envisageable circumstances. We have already seen that with just two degrees of freedom, a not very detailed study of a specific yet simple case (the Frahm oscillator) involved quite a lot of work.

In this chapter, we shall restrict ourselves to establishing the general form of the response due to any external forces with any initial conditions. Because of the linearity of the system, this response is the superposition of the free state caused by the initial conditions, which is already known to us (sections 12.2 and 12.7), and the forced state stimulated by the external forces. The latter will be calculated by means of the Laplace transformation and the convolution integral.

In order to simplify the presentation, we shall take the dissipative systems directly into consideration by assuming that all the modes are oscillatory and distinct. As in chapter 12, it is advantageous to consider separately

· systems with real modes, which satisfy the Caughey condition and conservative systems which are specific cases;
· systems with complex modes (general case).

In the last part of this chapter, we shall give an introduction to the experimental modal analysis of real systems.

## 14.2 Dissipative systems with real modes

Let us go back to relation (10.1) which represents, in matrix form, the differential equations of a system in the forced state.

$$[M]\ \ddot{\vec{x}} + [C]\ \dot{\vec{x}} + [K]\ \vec{x} = \vec{f}(t)$$

The right hand side is a vector with $n$ components consisting of any external forces acting on the system.

The Caughey condition being assumed satisfied, the change of basis

$$\vec{x} = [B]\ \vec{q}$$

allows one, as we have seen in section 12.2, to decouple the above system of equations. One obtains in this way

$$\lceil M^\circ \rfloor\ \ddot{\vec{q}} + \lceil C^\circ \rfloor\ \dot{\vec{q}} + \lceil K^\circ \rfloor\ \vec{q} = [B]^T\ \vec{f}(t) \tag{14.1}$$

By adopting the change of notation

$$\vec{f}^\circ(t) = [B]^T\ \vec{f}(t) \tag{14.2}$$

one defines the vector $\vec{f}^\circ(t)$ of **modal forces** applied to the system.

Let us divide equation (14.1) by the modal mass matrix

$$\ddot{\vec{q}} + \lceil M^\circ \rfloor^{-1} \lceil C^\circ \rfloor\ \dot{\vec{q}} + \lceil M^\circ \rfloor^{-1} \lceil K^\circ \rfloor\ \vec{q} = \lceil M^\circ \rfloor^{-1}\ \vec{f}^\circ(t)$$

Let us next use the definitions (12.18)

$$\ddot{\vec{q}} + \lceil 2\ \Lambda \rfloor\ \dot{\vec{q}} + \lceil \Omega_0^2 \rfloor\ \vec{q} = \lceil M^\circ \rfloor^{-1}\ \vec{f}^\circ(t) \tag{14.3}$$

One obtains in this way a system comprising n independent equations, that is to say for $p = 1, 2, \ldots, n$

$$\ddot{q}_p + 2\ \lambda_p\ \dot{q}_p + \omega_{0p}^2\ q_p = \frac{1}{m_p^\circ}\ f^\circ(t) \tag{14.4}$$

These equations are analogous to those of an elementary oscillator in the forced state whose solution we established in chapter 6. The use of the Laplace transformation and the convolution integral (6.7) allows one to write the solution of (14.4) directly from (6.13). Assuming that the initial conditions are all zero, one obtains

$$q_p = \frac{1}{m_p^\circ \omega_p} \int_0^t f_p^\circ(t-u)\, e^{-\lambda_p u} \sin \omega_p u \, du \qquad p = 1, 2, \ldots, n \tag{14.5}$$

One can then return to the initial coordinates by means of the modal matrix

$$\vec{x} = [B]\, \vec{q} = \sum_p^n \vec{B}_p\, q_p$$

The solution of the forced state with zero initial conditions is therefore

$$\vec{x} = \sum_p^n \vec{B}_p \frac{1}{m_p^\circ \omega_p} \int_0^t f^\circ(t-u)\, e^{-\lambda_p u} \sin \omega_p u \, du \tag{14.6}$$

It is sufficient to add the above solution to the free state (12.31) for any initial conditions. By adopting the notation of section 12.3, the modal vectors $\vec{B}_p$ become $\vec{\beta}_p$ . On the other hand the pth modal force is equal to the scalar product

$$f_p^\circ(t) = \vec{B}_p^T\, \vec{f}(t) = \vec{\beta}_p^T\, \vec{f}(t) \tag{14.7}$$

The general solution of the forced state is then written

$$\begin{aligned} \vec{x} = \sum_p^n \frac{1}{m_p^\circ} \vec{\beta}_p \Big( \frac{1}{\omega_p} \int_0^t f_p^\circ(t-u)\, e^{-\lambda_p u} \sin \omega_p u \, du \; + \\ + e^{-\lambda_p t} \big( \vec{\beta}_p^T [M]\, \vec{X}_0 \cos \omega_p t + \frac{1}{\omega_p} \vec{\beta}_p^T [M]\, \{\vec{V}_0 + \lambda_p \vec{X}_0\} \sin \omega_p t \big) \Big) \end{aligned} \tag{14.8}$$

The general solution of the forced state of a conservative system is deduced easily from (14.8). In this case, the coefficients $\lambda_p$ are all zero and the natural angular frequencies $\omega_p$ become $\omega_{0p}$

$$\begin{cases} \lambda_p = 0 \\ \omega_p = \omega_{0p} = \dfrac{k_p^\circ}{m_p^\circ} \end{cases} \qquad p = 1, 2, \ldots, n \tag{14.9}$$

The solution (14.8) becomes

$$\vec{x} = \sum_p^n \frac{1}{m_p^\circ} \vec{\beta}_p \Big( \frac{1}{\omega_{0p}} \int_0^t f_p^\circ(t-u) \sin \omega_{0p} u \, du + \vec{\beta}_p^T [M] \vec{X}_0 \cos \omega_{0p} t + \frac{1}{\omega_{0p}} \vec{\beta}_p^T [M] \vec{V}_0 \, \text{xin} \, \omega_{0p} t \Big) \tag{14.10}$$

This result shows that the last two terms, due to the initial conditions $\vec{X}_0$ and $\vec{V}_0$, maintain themselves indefinitely in a conservative system. For such a system, there does not exist a steady state, strictly speaking, as we have already seen in the case of an elementary oscillator.

## 14.3 Dissipative systems in the general case

The Hamiltonian formulation of the differential equations has shown, when the Caughey condition is not satisfied, that the vector of external forces $\vec{p}(t)$, of order $2n$, must be written in the form (12.66) :

$$\vec{p}^T = \{\vec{0}^T \; \vec{f}^T(t)\}$$

The differential system to be solved is thus

$$[D] \dot{\vec{y}} + [G] \vec{y} = \vec{P}(t) \tag{14.11}$$

The change of basis (12.74) and (12.75) allows one to decouple the system above which becomes

$$[D^\circ] \dot{\vec{q}} + [G^\circ] \vec{q} = [B]^T \vec{P}(t) \tag{14.12}$$

The matrix $[B]$ being complex, the modal forces are also. Let us call $P^\circ(t)$ the vector of modal forces

$$\vec{P}^\circ(t) = [B]^T \vec{P}(t) \tag{14.13}$$

By dividing the system (14.12) by the diagonal matrix $[D^\circ]$ and by using the definition (12.79), one can write

$$\dot{\vec{q}} + [\Delta] \vec{q} = [D^\circ]^{-1} \vec{P}^\circ(t) \tag{14.14}$$

This system consists of $2n$ independent differential equations

$$\dot{q}_p + \delta_p q_p = \frac{1}{d_p^\circ} P_p^\circ(t) \qquad p = 1, 2, \ldots, 2n \tag{14.15}$$

With the initial conditions all zero, integration of these equations gives

$$q_p = \frac{1}{d_p^\circ} \int_0^t P_p^\circ(t-u)\, e^{-\delta_p u}\, du \tag{14.16}$$

Let us again use the notation of section (12.7), that is to say

$$\delta_p = \lambda_p + j\omega_p$$

$$\delta_{p+n} = \lambda_p - j\omega_p = \delta_p^*$$

The solution above takes the form

$$q_p = \frac{1}{d_p^\circ} \int_0^t P_p^\circ(t-u)\; e^{-(\lambda_p + j\omega_p)u}\, du \tag{14.17}$$

and consequently

$$q_{p+n} = q_p^* = \frac{1}{d_p^{\circ *}} \int_0^t P_p^{\circ *}(t-u)\; e^{-(\lambda_p - j\omega_p)u}\, du \tag{14.18}$$

Let us come back to the initial coordinates by means of the modal matrix [B]

$$\vec{y} = [B]\,\vec{q} = \sum_p^{2n} \vec{B}_p\, q_p$$

The modal vectors $\vec{B}_p$ are also complex conjugate pairs, it gives thus

$$\vec{y} = \sum_p^{n} (\vec{B}_p\, q_p + \vec{B}_p^{*}\, q_p^{*}) \tag{14.19}$$

Before continuing, it is convenient to adopt the notation

$$\left\{\begin{array}{l} \vec{B}_p = \{\beta_{\ell p}\, e^{j\psi_{\ell p}}\} \\ d_p^\circ = D_p^\circ\, e^{j\alpha_p} \\ P_p^\circ(t) = |P_p^\circ(t)|\, e^{j\theta_p(t)} \end{array}\right. \qquad \left\{\begin{array}{l} \vec{B}_p^{*} = \{\beta_{\ell p}\, e^{-j\psi_{\ell p}}\} \\ d_p^{\circ *} = D_p^\circ\, e^{-j\alpha_p} \\ P_p^{\circ *}(t) = |P_p^\circ(t)|\, e^{-j\theta_p(t)} \end{array}\right. \tag{14.20}$$

One can now introduce (14.17) and (14.18) into (14.19), taking account of (14.20). By writing $z = t-u$, in order to simplify things, the $\ell$-th component of the solution $\vec{y}$ is given by the expression, with $\ell = 1, 2, \ldots, 2n$,

$$y_\ell = \sum_p^n (\beta_{\ell p}\, e^{j\psi_{\ell p}} \frac{1}{D_p^\circ}\, e^{-j\alpha_p} \int_0^t |P_p^\circ(z)|\, e^{j\theta_p(z)}\, e^{-(\lambda_p+j\omega_p)u}\, du +$$

$$+ \beta_{\ell p}\, e^{-j\psi_{\ell p}} \frac{1}{D_p^\circ}\, e^{j\alpha_p} \int_0^t |P_p^\circ(z)|\, e^{-j\theta_p(z)}\, e^{-(\lambda_p-j\omega_p)u}\, du)$$

By grouping the terms, this gives

$$y_\ell = \sum_p^n \beta_{\ell p} \frac{1}{D_p^\circ} \int_0^t |P_p^\circ(z)|\, e^{-\lambda_p u}\, [e^{-j(\omega_p u-\psi_{\ell p}-\theta_p(z)+\alpha_p)} +$$

$$+ e^{j(\omega_p u-\psi_{\ell p}-\theta_p(z)+\alpha_p)}]\, du$$

then, by introducing the harmonic function in place of the exponentials

$$y_\ell = \sum_p^n \beta_{\ell p} \frac{2}{D_p^\circ} \int_0^t |P_p^\circ(z)|\, e^{-\lambda_p u} \cos(\omega_p u-\psi_{\ell p}-\theta_p(z)+\alpha_p)\, du \tag{14.21}$$

The solution can be put in the vector form

$$\vec{y} = \sum_p^n \{\beta_{\ell p} \frac{2}{D_p^\circ} \int_0^t |P_p^\circ(z)|\, e^{-\lambda_p u} \cos(\omega_p u-\psi_{\ell p}-\theta_p(z)+\alpha_p)\, du\} \tag{14.22}$$

By adopting the same change of index as that of paragraph 12.7, the vector displacement $\vec{x}$ of the system, made up from the last $n$ components of $\vec{y}$, is given by the expression

$$\vec{x} = \sum_p^n \{\beta_{ip} \frac{2}{D_p^\circ} \int_0^t |P_p^\circ(z)|\, e^{-\lambda_p u} \cos(\omega_p u-\psi_{ip}-\theta_p(z)+\alpha_p)\, du\} \tag{14.23}$$

The transient state stimulated by the initial conditions is given by the relation (12.120), which was established in section 12.8. It is sufficient to superimpose it on the above result in order to obtain the most general form of the forced state

$$\vec{x} = \sum_{p}^{n} \{\beta_{ip} \left[\frac{2}{D_p^\circ} \int_0^t |P_p^\circ(z)| \, e^{-\lambda_p u} \cos(\omega_p u - \psi_{ip} - \theta_p(z) + \alpha_p) \, du \right. \qquad (14.24)$$

$$\left. + X_0' \, e^{-\lambda_p t} \cos(\omega_p t - \psi_{ip} - \varphi_0)\right]\}$$

It is useful to recall that this relation concerns a discrete dissipative generalized oscillator, taking into account however the following restrictive hypotheses

· all the modes are oscillatory by nature,
· all the eigenvalues are separate.

If, as well as the oscillatory modes, there exist some critical or super-critical modes simultaneously, the study of the forced state is seriously complicated, without being of much interest in the general case however. It would be more useful, but that goes beyond the framework set for this chapter, to examine the risk of instability which can appear in the behaviour of the system when two eigenvalues are very close to each other. This phenomenon is called **instability from confusion of the eigenvalues.**

## 14.4 Introduction to experimental modal analysis

The main objective of experimental modal analysis is to determine the dynamic characteristics of a real structure, in other words the frequencies, modes and natural mode shapes as well as the modal dampings. This discipline has experienced a major expansion in recent years because of the miniaturization and increase in performance of computerized systems for data acquisition and digital signal processing. It is based on the measurement of the ratio between a specific excitation of the structure and the response (displacement, velocities, accelerations, ... ) that this excitation causes.

The structure studied (car body, frame of a machine, steel bridge, etc.) is continuous by nature, whereas the measurements are made at points. It is therefore necessary to render the system discrete conceptually. The discrete points (nodes) are determined by the choice of both the points for the application of exciting forces and the points for the measurements.

Let us assume that the forces are applied to the points where the measurements are made, in accordance with the usual practice. The number of degrees of freedom $n$ of the sub-divided system is then equal to the product of the number of points of measurement $m$ by the number of generalized coordinates $r$ chosen for each of these points

$$n = m \cdot r \tag{14.25}$$

With such a procedure, it is clear that the mass, damping and stiffness matrices are not known a priori. Therefore, in order to determine the modal parameters, it is necessary to proceed in a different way from the preceding chapters, while using the results already established.

Let us come back to relation (14.11) which gives, in matrix form, the differential equations of the forced state in the general case, that is to say when the Caughey condition is not satisfied.

$$[D]\,\dot{\vec{y}} + [G]\,\vec{y} = \vec{P}(t)$$

Let us take the Laplace transform of the two sides by assuming the initial conditions are all zero.

$$[s[D] + [G]]\,\vec{Y}(s) = \vec{P}(s) \tag{14.26}$$

The square matrix of order $2\,n$ of the left-hand side is called the **operational impedance matrix**

$$[Z(s)] = [[G] + s[D]] \tag{14.27}$$

Its inverse, the **operational admittance matrix,** is also called the **matrix of transfer functions,**

$$[H'(s)] = [Z(s)]^{-1} \tag{14.28}$$

We are going to show that knowledge of the matrix $[H'(s)]$ is sufficient to determine all the eigenvectors of the system.

Let us rewrite the differential system (14.26) in the form

$$[Z(s)]\,\vec{Y}(s) = \vec{P}(s) \tag{14.29}$$

By referring to chapter 12, one sees that the characteristic equation of the homogeneous system is

$$\left|Z(s)\right| = 0 \tag{14.30}$$

It has $2\,n$ complex solutions, which are conjugate pairs, designated $\delta_1, \delta_2, \ldots, \delta_{2n}$. One can thus write it in the form

$$(s - \delta_1)\,(s - \delta_2)\,\ldots\,(s - \delta_{2n}) = 0$$

or, by using the product symbol $\Pi$,

$$\prod_{\ell}^{2n} (s - \delta_\ell) = 0 \tag{14.31}$$

As before, the eigenvectors are obtained by solving the homogeneous system

$$[Z(s{=}\delta_p)]\,\vec{B}_p = \vec{0} \tag{14.32}$$

The quantity $\delta_p$ defined above corresponds to that of chapter 12, apart from the sign. Using (12.89), it has the value

$$\delta_p = -\,\frac{g_p^\circ}{d_p^\circ} \tag{14.33}$$

If one designates the adjoint matrix of $[Z(s)]^a$ by $[Z(s)]$, the calculation of $[H'(s)]$ gives

$$[H'(s)] = [Z(s)]^{-1} = \frac{[Z(s)]^a}{|Z(s)|} \tag{14.34}$$

By using the form (14.31) of the characteristic equation in order to express the denominator it gives, A being a real constant,

$$[H'(s)] = \frac{[Z(s)]^a}{A \prod_{\ell}^{2n} (s-\delta_\ell)} \tag{14.35}$$

Since the values of $\delta_\ell$ are conjugate pairs, one can expand $[H'(s)]$ into a sum of simple elements of the form

$$[H'(s)] = \sum_{p}^{2n} \frac{[R'^p]}{s - \delta_p} = \sum_{p}^{n} \left(\frac{[R'^p]}{s - \delta_p} + \frac{[R'^{p^*}]}{s - \delta_p^*}\right) \tag{14.36}$$

in which $[R'^p]$ is called the **matrix of residues at the pole** $\boldsymbol{\delta_p}$ . This matrix is given by the expression

$$[R'^p] = \frac{[Z(s=\delta_p)]^a}{A \prod_{\ell \neq p}^{2n} (\delta_p-\delta_\ell)} \tag{14.37}$$

Let us put relation (14.34) into the equivalent form

$$[Z(s)]\ [Z(s)]^a = |Z(s)|\ [I] \tag{14.38}$$

The determinant $Z(s)$ is zero for each eigenvalue $s = \delta_p$ . One can thus write

$$[Z(s=\delta_p)]\ [Z(s=\delta_p)]^a = [0] \tag{14.39}$$

This relation implies, for the $2n$ column vectors of the matrix $[Z(s=\delta_p)]^a$

$$[Z(s=\delta_p)]\ \vec{Z}_k^a\ (s=\delta_p) = \vec{0} \tag{14.40}$$

The equations (14.32) and (14.40) are identical. Consequently, at the eigenvalue $\delta_p$, all the vectors $\vec{Z}_k^a(s=\delta_p)$ are equal, apart from a factor, to the eigenvector $\vec{B}_p$. On the other hand, the relationship (14.37) shows that the matrix of residues at the pole $[R'^p]$ is proportional to the matrix $[Z(s=\delta_p)]^a$. Thus the column vectors of $[R'^p]$ are also proportional to the eigenvector $\vec{B}_p$ for the eigenvalue $s = \delta_p$. This result shows that knowledge of the matrix $[H'(s)]$ is sufficient to determine the eigenvectors of the system.

Finally, let us also point out that the matrices $[Z(s)]$ and $[H'(s)]$ are symmetrical since the matrices $[D]$ and $[G]$ are.

We are now going to express the matrix of transfer functions $[H'(s)]$ independently of the matrices $[D]$ and $[G]$ which are not known in the present case. In order to do that, let us expand the vector $\vec{y}(t)$ in the modal basis formed from the vectors $\vec{B}_p$. Equation (12.107) allows one to write

$$\vec{y}(t) = \sum_p^{2n} \gamma_p(t)\ \vec{B}_p \tag{14.41}$$

By taking the Laplace transform of the two sides, it gives

$$\vec{Y}(s) = \sum_p^{2n} \Gamma_p(s)\ \vec{B}_p \tag{14.42}$$

Let us introduce this value into equation (14.26) after having expanded the product

$$s \sum_p^{2n} \Gamma_p(s)\ [D]\ \vec{B}_p + \sum_p^{2n} \Gamma_p(s)\ [G]\ \vec{B}_p = \vec{P}(s)$$

Pre-multiplied by the transpose of any eigenvector $\vec{B}_r$, the preceding equation becomes

$$s \sum_{p}^{2n} \Gamma_p(s)\, \vec{B}_r^T[D]\, \vec{B}_p + \sum_{p}^{2n} \Gamma_p(s)\, \vec{B}_r^T[G]\, \vec{B}_p = \vec{B}_r^T\, \vec{P}(s)$$

The orthogonality relations for the modal vectors (12.109) and (12.110) eliminate all the terms for which $r \neq p$ . It therefore reduces to

$$s\, \Gamma_r(s)\, d_r^{\circ} + \Gamma_r(s)\, g_r^{\circ} = \vec{B}_r^T\, \vec{P}(s) \tag{14.43}$$

One obtains in this way the value of $\Gamma_r(s)$

$$\Gamma_r(s) = \frac{\vec{B}_r^T\, \vec{P}(s)}{s\, d_r^{\circ} + g_r^{\circ}} \tag{14.44}$$

which becomes, by using the relation (14.33)

$$\Gamma_r(s) = \frac{\vec{B}_r^T\, \vec{P}(s)}{d_r^{\circ}(s - \delta_r)} \tag{14.45}$$

Taking account of the last result, relation (14.42) takes the form

$$\vec{Y}(s) = \sum_{p}^{2n} \frac{\vec{B}_p^T\, \vec{P}(s)\, \vec{B}_p}{d_p^{\circ}(s - \delta_p)} \tag{14.46}$$

The eigenvalues and eigenvectors being complex conjugate pairs, the terms of the sum above can be grouped as follows

$$\vec{Y}(s) = \sum_{p}^{n} \left(\frac{\vec{B}_p^T\, \vec{P}(s)\, \vec{B}_p}{d_p^{\circ}(s - \delta_p)} + \frac{\vec{B}_p^{*T}\, \vec{P}(s)\, \vec{B}_p^{*}}{d_p^{\circ *}(s - \delta_p^{*})}\right) \tag{14.47}$$

By definition, the vector P(s) is given by the expression

$$\vec{P}(s) = \{\vec{0}^T \ \vec{f}(s)^T\}$$

It is possible to express the last n components of $\vec{Y}(s)$ , that is to say the vector $\vec{X}(s)$ , in the form analogous to (14.47). Let us designate the vector form of the last n components of $\vec{B}_p$ by $\vec{b}_p$.

It gives

$$\vec{X}(s) = \sum_p^n \left(\frac{\vec{b}_p^T \ \vec{f}(s) \ \vec{b}_p}{d_p^\circ(s - \delta_p)} + \frac{\vec{b}_p^{*T} \ \vec{f}(s) \ \vec{b}_p^*}{d_p^{\circ *}(s - \delta_p^*)}\right) \tag{14.48}$$

This relation gives the response of the system to any excitation $\vec{f}(t)$ in the Laplace domain.

Let us assume that the vector $\vec{f}(t)$ has only a single non-zero component, the kth for example, and that this component consists of a unit Dirac delta function. The Laplace transform of $\vec{f}(s)$ and $\vec{f}(t)$ will then consist of only zero components, except for the kth, which will be equal to one. In practice, one way to do this is to supply an impulse, by means of a calibrated hammer blow, which is in the direction of the displacement $x_k$ .

By designating the response of the system to this impulse by $\vec{X}_k(s)$ and the kth component of the eigenvector $\vec{b}_p$ by $b_{pk}$ , one obtains

$$\vec{X}_k(s) = \sum_p^n \left(\frac{b_{pk} \ \vec{b}_p}{d_p^\circ(s - \delta_p)} + \frac{b_{pk}^* \ \vec{b}_p^*}{d_p^{\circ *}(s - \delta_p^*)}\right) \tag{14.49}$$

As we have seen for the elementary oscillator, the response of the system to a Dirac impulse is equal to the transfer function expressed by means of the Laplace transformation.

The vector $\vec{X}_k(s)$ defined by (14.49) is thus the kth column vector of the matrix of transfer functions [H(s)] , the latter consisting of the lower right sub-matrix of [H'(s)] and having dimensions n x n. Let us then designate the ith component of

$\vec{X}_k(s)$ by $H_{ik}(s)$ . This represents the transfer function between the generalized coordinates for rows $i$ and $k$

$$H_{ik}(s) = \sum_p^n \left(\frac{b_{pk}\, b_{pi}}{d_p^\circ (s - \delta_p)} + \frac{b_{pk}^*\, b_{pi}^*}{d_p^{\circ *}(s - \delta_p^*)}\right) \qquad (14.50)$$

With the writing simplified

$$R_{ik}^p = \frac{b_{pk}\, b_{pi}}{d_p^\circ} \qquad (14.51)$$

the relation (14.50) becomes

$$H_{ik}(s) = \sum_p^n \left(\frac{R_{ik}^p}{s - \delta_p} + \frac{R_{ik}^{p^*}}{s - \delta_p^*}\right) \qquad (14.52)$$

Let us come back to the matrix notation by introducing the matrices $[R^p]$ and $[R^{p^*}]$ , of order $n$ . The matrix $[H(s)]$ of the transfer functions, also of order $n$ , is consequently given by the expression

$$[H(s)] = \sum_p^n \left(\frac{[R^p]}{s - \delta_p} + \frac{[R^{p^*}]}{s - \delta_p^*}\right) \qquad (14.53)$$

The comparison of the relations (14.53) and (14.36) shows that the matrices $[R^p]$ consist of the lower right sub-matrices of the matrices $[R'^p]$. The terms $R_{ik}^p$ are therefore the residues at the pole $\delta_p$ and are equal, apart from a factor, to the components $b_{pi}$ of the eigenvector $\vec{b}_p$ .

The matrix $[H'(s)]$ being symmetrical, the matrix $[H(s)]$ is also. On the other hand, for a value $s = \delta_p$ , all the columns of $[H(s)]$ are proportional to the eigenvector $\vec{b}_p$. As a result, knowledge of a single line or of a single column of $[H(s)]$ is sufficient to determine all of the $n$ eigenvectors $\vec{b}_p$, taking into account the results (14.36) and (14.37).

Furthermore, the matrix of transfer functions $[H(s)]$ allows one to find the response of the system to any excitation, by means of the relation

$$\vec{X}(s) = [H(s)]\ \vec{f}(s) \tag{14.54}$$

In principle, in this way one can determine a column of $[H(s)]$ by giving an impulse to a single degree of freedom of the structure and then by measuring the responses for all the degrees of freedom initially chosen. Conversely, it is possible to determine a row of $[H(s)]$ by measuring the response of the structure for a single degree of freedom for impulses supplied to all the degrees of freedom.

In practice, this procedure is less simple because the digital signal analyzers which are actually available are based on the discrete Fourier transform (FFT analyzer). They allow one to establish, not the transfer functions $H_{ik}(s)$ of the complex variable $s$ of Laplace, but only the frequency responses $H_{ik}(\omega)$ whose analytical form can be obtained by replacing $s$ by $j\omega$ in $H_{ik}(s)$.

This situation is illustrated in figure 14.1. The real and imaginary parts of $H_{ik}(s)$ describe surfaces whose intersections with the vertical plane 023 and 023' represent the real and imaginary parts of $H_{ik}(\omega)$ respectively.

The fact that only the frequency response $H(\omega)$ is accessible experimentally means a loss of information which makes determination of the modal dampings difficult. Nevertheless, the storage, display and processing capacities of actual computerized analyzers for the frequency responses makes it possible, by means of an adequate procedure, to obtain the modal quantities with a precision and a reliability which is quite sufficient in the majority of cases.

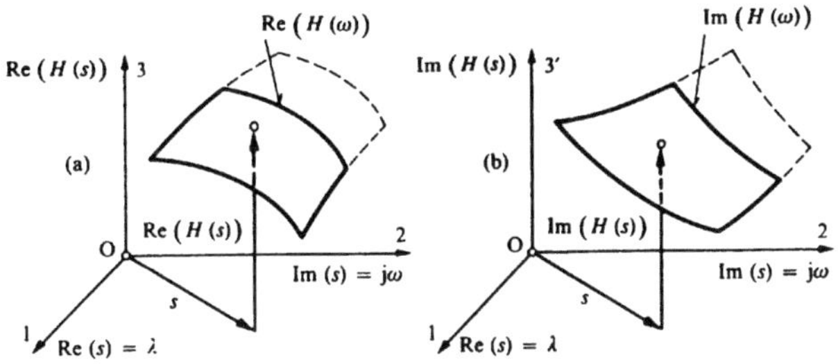


Fig. 14.1 Transfer function $H(s) = H_{ik}(s)$ between the generalized coordinates of rows i and k

(a) Real part of $H(s)$ ; the intersection of the surface with the plane 023 gives the real part of the frequency response $H(\omega)$
(b) Imaginary part of $H(s)$ ; the intersection of the surface with the plane 023' gives the imaginary part of $H(\omega)$

We shall not undertake the study of this procedure here, nor that of the algorithms which are associated with it, for our objective is limited to the presentation of the principles of experimental modal analysis. We shall confine ourselves to point out that the operator selects the number of modes and their approximate frequencies, for a specific range of frequencies and on the basis of a visual examination of one or more of the frequency responses. Then some appropriate algorithms allow one to find a new approximation for the modal frequencies as well as a first approximation for the modal dampings. Then curve-fitting techniques allow one to determine the eigenvalues and vectors. In this way, for each frequency response $H_{ik}(\omega)$ of a system having n degrees of freedom, it is necessary to adjust $4n + 2$ parameters, to know

- 2 n parameters for the n complex eigenvalues
- 2 n parameters for the pth components of each of the n complex eigenvectors
- 2 parameters for the two terms, one of which takes into account the modes situated below the frequency band chosen for analysis, and the other for the modes situated above it.

Once the various modal parameters are determined in this way, the behaviour of the structure can be describe in a convenient way.

# BIBLIOGRAPHY


[1] Caughey, T.K., O'Kelly, M.E.J. Classical Normal Modes in Damped Linear Dynamic Systems, J. Appl. Mech., sept. 1965, 583-588

[2] Clough, R.W., Penzien, J. Dynamic of Structures, McGraw-Hill, 1975

[3] Den Hartog, J.P. Vibrations mécaniques (2nd ed.), Dunod, Paris, 1960

[4] Dimentberg, F.M. Flexural Vibrations of Rotating Shafts, Butterworth, London, 1961

[5] Fertis, D.G. Dynamics and Vibrations of Structures, Krieger, Melbourne FL, 1983

[6] Filipov, A.P. Vibrations of Mechanical Systems, (translated by Z.F. Reif), National Lending library, Boston Spa, 1971

[7] Frazer, R.A., Duncan, W.J., Collar, A.R. Elementary Matrices, Cambridge Univ. Press. N.Y., 1957

[8] Harker, R.J. Generalized Methods of Vibration Analysis, J. Wiley and Sons, N.Y., 1983

[9] Harris, C.M., Crede, C.E. Shock and Vibration Handbook (2nd ed. 3 vol.), McGraw Hill, 1976

[10] Hurty, W.C., Rubinstein, M.F. Dynamics of Structures Prentice-Hall, Englewood Cliffs, N. J., 1964

[11] Klostermann, A.L. On the Experimental Determination and Use of Modal Representations of Dynamic Characteristics, University of Cincinnati, Thesis, 1971

[12] Klotter, K. Technische Schwingungslehre (2 vol.), Springer, Berlin, Heidelberg, N.Y., 1980-81

[13] Kolsky, H. Stress Waves in Solids, Dover, N.A., 1963

[14] Lalanne, M., Berthier, P., Der Hagopian, J. Mécanique des vibrations linéaires (2nd ed.), Masson, Paris, 1986

[15] Lur'é, L. Mécanique analytique, (volumes I and II) Masson, Paris, 1968

[16] Mathey, R. Physique des vibrations mécaniques, Dunod, Paris, 1963

[17] Mazet, R. Mécanique vibratoire (2nd ed.), Dunod, Paris, 1966

[18] Meirovitch, L. Analytical Methods in Vibrations, Macmillan Company, N.Y., 1967

[19] Meirovitch, L. Elements of Vibration Analysis, McGraw-Hill, 1975

[20] Müller, P.C. Stabilität und Matrizen Springer. Verlag, Berlin, 1977

[21] Nashif, A.D., Jones, D.I.G., Henderson, J.P. Vibration Damping, J. Wiley and Sons, N.Y., 1985

[22] Parlett, B.N. The Symmetric Eigenvalue Problem, Prentice-Hall, 1980

[23] Preumont, A. Vibrations aléatoires et analyse spectrale, Presses polytechniques romandes, Lausanne 1988

[24] Rao J.S., Gupta, K. Theory and Practice of Mechanical Vibrations, Wiley East, Ltd, 1984

[25] Rayleigh, Lord — The Theory of Sound, Vol. 1, Dover, N.Y., 1945

[26] Rocard, Y. — Dynamique générale des vibrations (4th ed.), Masson, Paris, 1971

[27] Roseau, M. — Vibrations des systèmes mécaniques, Masson, Paris, 1984

[28] Salles, F., Lesueur, C. — Les vibrations mécaniques, ed. Masson, Paris, 1972

[29] Sneddon, I.N. — Fourier Transforms, McGraw-Hill, N.Y., 1951

[30] Soutif, M. — Vibrations, propagation, diffusion, Dunod, Paris, 1970

[31] Steidel, R.F. — An Introduction to Mechanical Vibrations, (2nd ed.), J. Wiley and Sons, 1980

[32] Thomson, W.T. — Theory of Vibrations with Applications (2nd ed.), Prentice-Hall, 1981

[33] Timoshenko, S., Young, D.H., Weaver, W. — Vibration Problems in Engineering (4th ed.), J. Wiley and Sons, N.Y., 1974

[34] Tse, F.S., Morse, I.E., Minkle, R.T. — Mechanical Vibrations (2nd ed.), Allyn & Bacon, 1978

[35] Wylie, R.C., Barrett, L.C. — Advanced Engineering Mathematics (5th ed.), McMillan, 1982

# I N D E X

· The numbers which follow each entry refer to the corresponding pages in the book.

· When several numbers are given, the main reference(s) is (are) indicated by bold type.

· When it helps one to understand the terms in the Index, the nature of the system is indicated by means of the following abbreviations :

02 : oscillator with two degrees of freedom

OG : generalized oscillator with n degrees of freedom

## Proper Names

Subject

# SYMBOL LIST

- We have used latin and greek letters only and we have resorted to the least possible number of indices. Consequently, it is inevitable that certain symbols have several meanings. Thus, for example, T represents the kinetic energy of a system, the tension in a wire or the period of an oscillating quantity. However, confusion is highly unlikely.

- Symbols appearing only briefly, in particular in the applied examples, are not mentioned in this list.

- When it appears to be necessary to identify the symbols below more clearly (but not in an exhaustive way), the following abbreviations are added :

  O2 : oscillator with two degrees of freedom
  OG : generalized oscillator (with n degrees of freedom)
  OGA : conservative generalized oscillator
  OGB : dissipative generalized oscillator with real modes (ie the Caughey condition is satisfied)
  OGC : dissipative generalized oscillator with complex modes (ie the general case where the Caughey condition is not satisfied).

## General symbols

It is useful to define the meaning, in this book, of several more general symbols.

$u(t)$ function of time

$\dot{u}(t)$ first derivative of $u(t)$ with respect to time

$u(t)$ second derivative of $u(t)$ with respect to time

$\vec{v} = \begin{Bmatrix} v_1 \\ \vdots \\ v_n \end{Bmatrix}$ vector having n components $v_i$

$\vec{v}^T = (v_1 \dots v_n)$ transposed vector of $\vec{v}$

$\|\vec{v}\|$ magnitude of the vector $\vec{v}$

$j = \sqrt{-1}$ symbol for the square root of minus one

$\underline{N}$ complex number

$\underline{N}^*$ complex conjugate of $\underline{N}$

Re $\underline{N}$ real part of the complex number $\underline{N}$

Im $\underline{N}$ imaginary part of the complex number $\underline{N}$

<u>Note</u> : complex numbers are no longer underlined from section 12.4, page 189

$i, j, k, l, m, n, r, s, t, \alpha$ indices, whole numbers

$\pi$ ratio of the circumference to the radius (= 3,1416)

$\Pi$ symbol for a product

$\Sigma$ symbol for a summation

$[M]$ square matrix M of order n

$[M]^a$ adjoint matrix of $[M]$

$\lceil M^{\circ} \rfloor$ diagonal matrix of order n

**Latin alphabet**

a speed of propagation of elastic waves in a solid

a real part of the complex frequency response

A area of a section

| | |
|---|---|
| A | real constant (in particular, the constant of integration) |
| $A_n$ | coefficient of the nth cosine term of a Fourier series |
| $\underline{A}$ | complex constant |
| [A] | core matrix (OGA, order n) |
| b | imaginary part of the frequency response |
| $\vec{b}$ | eigenvector |
| B | real constant (in particular, the constant of integration) |
| $B_n$ | coefficient of the nth sine term of a Fourier series |
| $\underline{B}$ | complex constant |
| $\vec{B}_p$ | eigenvector of the matrix [B] (OGA, B, order n; OGC, order 2n) |
| [B] | change of basis matrix (OGA, B, order n; OGC, order 2n) |
| c | linear viscous damping factor (or linear resistance) |
| c' | critical damping factor |
| $c_j$ | jth resistance acting on a point mass (coplanar system, O2) |
| $c_p^o$ | modal damping factor of the pth natural mode (OGB) |
| C | electrical capacity |
| $C_o$ | half the mean value of a periodic force (=1/2 $F_o$) |
| $\underline{C}_n$ | complex amplitude of the term $e^{jn\omega t}$ of a complex Fourier series ($\underline{C}_n^*$ = conjugate of $\underline{C}_n$) |
| [C] | damping matrix (or loss matrix) (OG, order n) |
| $\lceil C^o \rfloor$ | diagonal modal damping matrix (OGB, order n) |

| | |
|---|---|
| $d(t)$ | impulse response (response of an oscillator to an impulse excitation) |
| $d_p^o$ | complex term of the pth row of the diagonal matrix $\lceil D^{o} \rfloor$ ($d_r^o$ , same but the rth row) |
| $D_p^o$ | modulus of the complex term $d_p^o$ |
| $D(s)$ | Laplace transform of Dirac delta function (=1) |
| $\underline{D}_n$ | complex amplitude of the nth harmonic of a complex Fourier series |
| $[D]$ | matrix for a dissipative oscillator (OGC, order 2n) |
| $\lceil D^{o} \rfloor$ | diagonal matrix for a dissipative oscillator (OGC, order 2n) |
| | |
| $e(t)$ | indicial response (response of an oscillator to a unit step force) |
| $E$ | modulus of elasticity (Young's modulus) |
| $E(s)$ | Laplace transform of a step function |
| $E_s$ | static displacement caused by a step function |
| $[E]$ | inverse core matrix (= $[A]^{-1}$, OGA, order n) |
| | |
| $f$ | frequency of oscillating quantity ($f = 2\pi/\omega$ , see letter $\omega$) |
| $f(t)$ | external force applied to a system; function of time |
| $f_c(t)$ | viscous frictional force |
| $f_i(t)$ | external force associated with the generalized coordinate $x_i(t)$ |
| $f_k(t)$ | elastic restoring force |
| $f_m(t)$ | force of inertia |
| $\underline{f}(t)$ | complex external force |
| $\vec{f}(t)$ | vector of external forces applied to a system |
| $\vec{f}^{o}(t)$ | vector of modal forces (OGA,B order n) |
| $F$ | amplitude of an harmonic force |
| $F(s)$ | Laplace transform of a function $f(t)$ |

| | |
|---|---|
| $F'(s)$ | Carson-Laplace transform of a function f(t) |
| $F_n$ | amplitude of the nth harmonic of a periodic force |
| $F_o$ | mean value of a periodic force |
| $\underline{F}(\omega)$ | Fourier transform of a function f(t) |
| [F] | core matrix for a dissipative oscillator (OGC, order 2n) |
| | |
| g | terrestrial gravitational constant (= 9.81 $m/s^2$) |
| g(t) | response of an oscillator to a step displacement |
| $g_r^o$ | term of the nth row of the diagonal matrix $\lceil G^o \rfloor$ |
| [G] | matrix for a dissipative oscillator (OGC, order 2n) |
| $\lceil G^o \rfloor$ | diagonal matrix for a dissipative oscillator (OGC, order 2n) |
| | |
| h(t) | response of an oscillator to a unit impulse displacement |
| H | total energy of an oscillator (of a system) |
| H | Hamiltonian function (or Hamiltonian, = T+V) |
| H(s) | operational response of an oscillator |
| $H_d$ | energy lost during one period of a dissipative oscillator |
| $H_o$ | initial total energy of a dissipative oscillator |
| $H_{ik}(s)$ | transfer function between the ith and jth displacements |
| $H_{ik}(\omega)$ | frequency response between the ith and jth displacements |
| $\bar{H}$ | mean total energy of a dissipative oscillator |
| $\underline{H}$ | complex frequency response |
| $\underline{H}_o$ | complex response at the natural frequency of a conservative oscillator, as well as at the velocity, power and phase resonances |
| $\underline{H}_1$ | complex response at the natural frequency of a dissipative oscillator |
| $\underline{H}_2$ | complex response at the amplitude resonance |
| $\underline{H}_3$ | complex response at the acceleration resonance |
| $\underline{H}_n$ | complex frequency response of the nth harmonic to a periodic displacement |

| | |
|---|---|
| [H(s)] | transfer function matrix (OGC, order n) |
| [H'(s)] | expanded matrix of transfer functions (or operational admittance matrix) (OGC, order 2n) |
| i(t) | electric current |
| I | amplitude of an alternating electric current |
| I | moment of inertia of a plane area about an axis (bending moment of inertia) |
| $I_p$ | polar moment of inertia of a plane area (torsional moment of inertia) |
| $\lceil I \rfloor$ | unit matrix (or identity matrix) |
| J | moment of inertia of a solid (mass moment of inertia) |
| k | stiffness of an elastic element |
| $k_e$ | equivalent stiffness of an elastic system |
| $k_i$ | stiffness of the ith spring acting on a point mass (coplanar system, O2) |
| $k_{ij}$ | coupling stiffness between the generalized coordinates $x_i$ and $x_j$ (= $k_{ji}$) |
| $k_p^o$ | modal stiffness of the pth natural mode ($k_r$, same but rth mode) |
| [K] | stiffness matrix (OG, order n) |
| $\lceil K^o \rfloor$ | diagonal modal stiffness matrix (OGA,B, order n) |
| L | distance, length of a beam |
| L | self-inductance |
| L | Lagrange function (or Lagrangian), (= T-V) |
| L(f(t)) | Laplace transform of a function f(t), (= F(s)) |

| | |
|---|---|
| $m$ | mass of an oscillator (of a system) |
| $m_{ij}$ | generalized mass (or coefficient of inertia), ($= m_{ji}$) |
| $m_\alpha$ | $\alpha$th mass of a system of point masses |
| $[M]$ | mass matrix (or inertia matrix) (OG, order n) |
| $\lceil M^o \rfloor$ | diagonal modal mass matrix (OGA, B, order n) |
| $\|M\|$ | determinant of the mass matrix |
| $n$ | speed of rotation (in revolutions per minute) |
| $p$ | pressure |
| $p$ | root of the characteristic equation (O2) |
| $p(t)$ | instantaneous power consumed by an oscillator, in the natural or steady state |
| $p_k(t)$ | generalized motion variable |
| $p_p^o(t)$ | pth complex modal force |
| $\bar{p}$ | mean power consumed by an oscillator in a steady state |
| $\bar{p}_o$ | mean power consumed by an oscillator in a steady state, at the power resonance |
| $\vec{p}(t)$ | vector of generalized variables of motion |
| $\vec{p}(t)$ | vector of external forces (OGC, order 2n) |
| $\vec{p}^o(t)$ | vector of modal forces (OGC, order 2n) |
| $\vec{p}(s)$ | vector of Laplace transforms of the external forces (OGC, order 2n) |
| $q_p(t)$ | pth normal (or modal, or decoupled) coordinate |
| $\vec{q}(t)$ | vector of normal coordinates (OGA, B, order n; OGC, order 2n) |
| $Q$ | quality factor of an oscillator |

| | |
|---|---|
| $Q_i$ | generalized force acting on an elastic system |
| $Q_p$ | amplitude of the harmonic function $q_p(t)$ |
| $Q_p(s)$ | Laplace transform of $q_p(t)$ |
| $\vec{Q}$ | vector of generalized forces $Q_i$ acting on an elastic system |
| | |
| r | root of a characteristic equation |
| R | gas constant |
| R | radius of a circle (of a cylinder, of a pulley, etc.) |
| R | electric resistance |
| R(s) | radius of convergence of a Laplace transform |
| R(t) | radius of the phase plane spiral |
| $R(\vec{u})$ | Rayleigh quotient |
| $R^p_{ik}$ | residue at the pole $\delta_p$ |
| $[R^p]$ | matrix of the residues at the pole $\delta_p$ (OGC, order n) |
| $[R'^p]$ | expanded matrix of the residues at the pole $\delta_p$ (OGC, order 2n) |
| | |
| s | Laplace variable |
| $s_{ij}$ | element of a symmetric matrix ($= s_{ji}$) |
| [S] | symmetric matrix |
| | |
| t | time |
| T | kinetic energy of an oscillator (of a system) |
| T | tension in a wire or cord |
| T | period of an oscillating quantity ($= 2\pi/\omega$, see letter $\omega$) |
| $T_o$ | period of a conservative oscillator |
| $T_1$ | period of a dissipative oscillator |
| | |
| u | auxiliary variable of a convolution integral |
| u | electric tension |

$\vec{u}$ approximate eigenvector (OG, order n)

U amplitude of an alternating electric voltage

[U] inverse core matrix of a dissipative oscillator ($=[F]^{-1}$, OGC, order 2n)

v velocity

$v_\alpha$ velocity of a point mass $m_\alpha$

V volume

V amplitude of an harmonic velocity

V potential energy of an oscillator (of a system)

$V_o$ initial velocity

$V_\infty$ potential energy transmitted to an oscillator by a unit-step force (indicial response)

$\vec{V}_o$ vector of initial velocities (OG, order n)

$w_p$ inverse of the eigenvalue $\delta_p$

W dissipation function (half the power dissipated)

x(t) displacement of an oscillator

$\dot{x}(t)$ velocity

$\ddot{x}(t)$ acceleration

$x_1(t)$, $x_2(t)$ generalized displacement of an oscillator with 2 degrees of freedom (including the Frahm damper)

$x_a(t)$ displacement due to an external force (inverse of Y(s) F(s))

$x_b(t)$ displacement corresponding to the initial conditions $X_o$, $V_o$

$x_e(t)$ elastic displacement

$x_i(t)$ ith generalized coordinate (generalized displacement)

$x_{ip}(t)$ ith generalized coordinate (generalized displacement) in the pth natural mode

| | |
|---|---|
| $x'(t)$ | specific solution (displacement) of the differential equation without right-hand side |
| $x''(t)$ | general solution (displacement) of the differential equation without right-hand side |
| $\underline{x}(t)$ | complex displacement |
| $\underline{x}_e(t)$ | complex elastic displacement |
| $\vec{x}(t)$ | vector of displacements (OG, order n) |
| $\dot{\vec{x}}(t)$ | vector of velocities (OG, order n) |
| $\ddot{\vec{x}}(t)$ | vector of accelerations (OG, order n) |
| $\vec{x}_p(t)$ | pth natural mode |
| $X$ | amplitude of an harmonic displacement |
| $X(s)$ | Laplace transform of the displacement $x(t)$ |
| $X_0$ | initial displacement |
| $X_1$, $X_2$ | amplitudes of the harmonic displacements of an oscillator with 2 degrees of freedom (including the Frahm damper) |
| $X_p$ | reference amplitude for the pth natural mode |
| $X_n$ | amplitude of the nth harmonic of a periodic displacement |
| $X_s$ | static displacement |
| $X_{sn}$ | static displacement of the nth harmonic of a periodic displacement |
| $X'$ | reference amplitude of the mode $\vec{x}_r(t)$ caused by the initial conditions $\vec{X}_0$, $\vec{V}_0$ (OGC) |
| $X_e(s)$ | Laplace transform of the elastic displacement $x_e(t)$ |
| $\underline{X}$ | complex amplitude of the displacement |
| $\underline{X}(\omega)$ | Fourier transform of the displacement $x(t)$ |
| $\underline{X}_{sn}$ | complex static displacement of the nth harmonic of a periodic displacement |
| $\vec{X}(s)$ | vector of Laplace transforms of the displacements (OG, order n) |

| | |
|---|---|
| $\vec{X}_k(s)$ | vector of the Laplace transforms of the displacements $x_{ik}(t)$ due to an impulse following the displacement $x_k$ (OG, order n) |
| $\vec{X}_o$ | vector of initial displacements (OG, order n) |
| $\vec{X}_p$ | pth natural mode shape (vector of amplitudes of the pth natural mode( )OGA, B, order n) |
| $y(t)$ | temporal admittance |
| $\vec{y}(t)$ | vector of velocities and displacements (OGC, order 2n) |
| $Y(s)$ | operational admittance |
| $\underline{Y}$ | complex admittance |
| $\underline{Y}_n$ | complex admittance of the nth harmonic of a periodic displacement |
| $\vec{Y}_o$ | vector of initial velocities and of initial displacements (OGC, order 2n) |
| $\vec{Y}(s)$ | vector of the Laplace transforms of the velocities and displacements (OGC, order 2n) |
| $z$ | auxiliary variable to calculate convolution integrals (= t-u) |
| $Z(s)$ | operational impedance |
| $z_\alpha^k$ | cartesian coordinate of a point mass $m_\alpha$ |
| $\underline{Z}$ | complex impedance |
| $\vec{Z}^a_k$ | column vector of the adjoint matrix $[Z(s=\delta_p)]$, (OGC, order 2n) |
| $[Z(s)]$ | operational impedance matrix (OGC, order 2n) |

**Greek alphabet**

| | |
|---|---|
| $\alpha$ | angle, phase, phase shift |

| | |
|---|---|
| $\alpha$ | ratio of the natural angular frequencies in a Frahm damper ($= \omega_2/\omega_1$) |
| $\alpha_i$ | any real coefficient |
| $\alpha_i$ | angle defining the direction of the ith spring acting on a point mass (coplanar system, 02) |
| $\alpha_{ij}$ | influence coefficient between the coordinates $x_i$ and $x_j$ ($= \alpha_{ji}$) |
| $\alpha_p$ | argument of the complex term $d_p^o$ |
| $[\alpha]$ | flexibility matrix (matrix of influence coefficients) |
| | |
| $\beta$ | angle, phase, phase shift |
| $\beta$ | relative angular frequency of an harmonic external force ($= \omega/\omega_0$) |
| $\beta_{ip}$ | relative amplitude of the coordinate $x_{ip}(t)$ of the pth natural mode ($\beta_{ir}$, the same but rth mode) |
| $\vec{\beta}_1$, $\vec{\beta}_2$ | eigenvectors of an oscillator with 2 degrees of freedom |
| $\vec{\beta}_p$ | pth natural mode shape (vector of the relative amplitudes of the pth natural mode), ($\vec{\beta}_r$, $\vec{\beta}_s$, the same but rth and sth modes) |
| | |
| $\gamma$ | isentropic exponent in the equation of state of a gas |
| $\gamma$ | angle, in the phase plane, between the tangent to the spiral and the normal to the radius |
| $\gamma_p$ | multiplicative constant, real or complex |
| $\vec{\gamma}$ | vector of the constant $\gamma_p$ |
| | |
| $\delta$ | deflection, static displacement |
| $\delta_p$ | pth eigenvalue |
| $\delta_{rs}$ | Kronecker delta ($= 1$ if $r = s$; $= 0$ if $r \neq s$) |
| $\lceil \Delta \rfloor$ | diagonal matrix of the eigenvalues $\delta_p$ (OGA,B order n; OGG, order 2n) |

| | |
|---|---|
| $\varepsilon$ | relative deformation (strain) in a material |
| $\varepsilon$ | relative power in a steady state (= $\bar{p}/\bar{p}_0$) |
| $\varepsilon$ | ratio of the masses in a Frahm damper (= $m_2/m_1$) |
| $\varepsilon_{max}$ | relative power at a power resonance |
| | |
| $\eta$ | damping factor (or relative damping) |
| $\eta_p$ | modal damping factor of the pth natural mode (or relative modal damping) |
| | |
| $\theta$ | angle, angle of rotation of a system |
| $\theta_p(t)$ | argument of the complex modal force $p^o(t)$ |
| $\theta'_p$, $\theta''_p$ | angles of the principal axes of the trajectory of a mass of a coplonar system of springs and linear resistances |
| $\lambda$ | damping coefficient |
| $\lambda_p$ | modal damping coefficient of the pth natural mode |
| $\Lambda$ | logarithmic decrement |
| $\lceil 2\Lambda \rfloor$ | diagonal modal damping matrix |
| | |
| $\mu$ | coefficient of dry friction (Coulomb friction) |
| $\mu$ | dynamic amplification factor |
| $\mu_1$ | linearly distributed mass (mass per unit length) |
| $\mu_0$ | dynamic amplification factor at the phase resonance |
| $\mu_{max}$ | dynamic amplification factor at the amplitude resonance |
| $\mu_n$ | dynamic amplification factor of the nth harmonic of a periodic displacement |

| | |
|---|---|
| $\xi_j$ | angle defining the direction of the jth resistance acting on a point mass (coplanar system) |
| $\rho$ | density (of matter) |
| $\sigma$ | normal stress (tensile, bending, ...) |
| $\tau$ | damping constant of a polymer |
| $\tau$ | frequency function (= $1/\delta$) |
| $\varphi$ | angle, phase shift between two oscillating quantities |
| $\varphi_1$, $\varphi_2$ | phase shifts of the displacements of an oscillator with two degrees of freedom |
| $\varphi_n$ | phase shift of the nth harmonic of a periodic displacement |
| $\varphi_p$ | phase shift of a pth normal mode (OG) |
| $\varphi_{or}$ | phase shift of the mode $\vec{x}_r(t)$ stimulated by the initial conditions $\vec{V}_o$, $\vec{X}_o$ (OGC) |
| $\Phi$ | magnetic flux |
| $\psi_n$ | phase shift of the nth harmonic of a periodic force |
| $\psi_{ip}$ | complementary phase shift of the displacement $x_{ip}(t)$ of a pth complex normal mode (OGC), ($\psi_{ir}$, the same but rth mode) |
| $\omega$ | angular frequency of an oscillating quantity |
| $\omega_0$ | natural angular frequency of a conservative oscillator<br>angular frequency at the phase resonance<br>angular frequency at the power resonance<br>angular frequency at the velocity resonance |

| | |
|---|---|
| $\omega_1$ | natural angular frequency of a dissipative oscillator |
| $\omega_2$ | angular frequency at the amplitude resonance |
| $\omega_3$ | angular frequency at the acceleration resonance |
| $\left.\begin{matrix}\omega_1\\ \omega_2\end{matrix}\right\}$ | natural angular frequencies of an oscillator with two degrees of freedom |
| $\omega_p$ | natural angular frequency of the pth mode |
| $\left.\begin{matrix}\Omega_1\\ \Omega_2\end{matrix}\right\}$ | angular frequencies for zero coupling for an oscillator with 2 degrees of freedom |
| $\Omega^4_{12}$ | term representing the elastic coupling of an oscillator with two degrees of freedom |
| $\lceil \Omega^2_0 \rfloor$ | diagonal matrix of the natural angular frequencies of an oscillator with real modes (OGA,B, order n) |

MIX
Papier aus verantwortungsvollen Quellen
Paper from responsible sources
FSC® C105338

If you have any concerns about our products,
you can contact us on
**ProductSafety@springernature.com**

In case Publisher is established outside the EU,
the EU authorized representative is:
**Springer Nature Customer Service Center GmbH**
**Europaplatz 3, 69115 Heidelberg, Germany**

Printed by Libri Plureos GmbH
in Hamburg, Germany